T0318823

Building Wireless Sensor Networks

Building Wireless Sensor Networks

Application to Routing and Data Diffusion

Edited by

Smain Femmam

ELSEVIER

First published 2017 in Great Britain and the United States by ISTE Press Ltd and Elsevier Ltd

ISTE Press Ltd
27-37 St George's Road
London SW19 4EU
UK

www.iste.co.uk

Elsevier Ltd
The Boulevard, Langford Lane
Kidlington, Oxford, OX5 1GB
UK

www.elsevier.com

Notices

Knowledge and best practice in this field are constantly changing. As new research and experience broaden our understanding, changes in research methods, professional practices, or medical treatment may become necessary.

Practitioners and researchers must always rely on their own experience and knowledge in evaluating and using any information, methods, compounds, or experiments described herein. In using such information or methods they should be mindful of their own safety and the safety of others, including parties for whom they have a professional responsibility.

To the fullest extent of the law, neither the Publisher nor the authors, contributors, or editors, assume any liability for any injury and/or damage to persons or property as a matter of products liability, negligence or otherwise, or from any use or operation of any methods, products, instructions, or ideas contained in the material herein.

MATLAB$^{®}$ is a trademark of The MathWorks, Inc. and is used with permission. The MathWorks does not warrant the accuracy of the text or exercises in this book. This book's use or discussion of MATLAB$^{®}$ software or related products does not constitute endorsement or sponsorship by The MathWorks of a particular pedagogical approach or particular use of the MATLAB$^{®}$ software.

For information on all our publications visit our website at http://store.elsevier.com/

British Library Cataloguing-in-Publication Data
A CIP record for this book is available from the British Library
Library of Congress Cataloging in Publication Data
A catalog record for this book is available from the Library of Congress
ISBN 978-1-78548-274-8

Printed and bound in the UK and US

Contents

Smain FEMMAM

Xiaoting LI and Laurent GEORGE

Chapter 2. Representation of Networks of Wireless Sensors with a Grayscale Image: Application to Routing   31

Marc GILG

Chapter 3. Routing and Data Diffusion in Vehicular Ad Hoc Networks   67

Frédéric DROUHIN and Sébastien BINDEL

Chapter 6. One-by-One Embedding of the Twisted Hypercube into Pancake Graph

Smain FEMMAM and Faouzi M. ZERARKA

Preface

The objective of this book is to enable both practitioners and researchers to gain a full understanding of the new approaches to secure routing in wireless sensor networks with different new hybrid topologies. Existing wireless networks aim to provide communication services between vehicles by enabling the vehicular networks to support wide range applications, for instance, enhancing the efficiency of road transportation.

We present an analysis of the security of real-time data diffusion, a protocol used for routing in wireless sensor networks. Along with this, we introduce the various possible attacks on this routing protocol and the possible countermeasures to prevent these attacks. Different applications are introduced, and different new topologies are developed for these purposes.

Several new approaches and their efficiency to resolve real problems encountered in the practical cases in wireless networks are explored.

Audio video bridging (AVB) switched Ethernet is being considered as a promising evolution of switched Ethernet networks for the transmission of time-sensitive frames. In this discussion, we present a survey of Ethernet AVB, focusing on Physical and MAC layers and on the latest development called time-sensitive networks (TSN). Then, we recall existing solutions to ensure real-time timeliness constraints for Ethernet AVB flows sent according to the credit-based shaping algorithm (CBSA). We present new bounds on the worst-case end-to-end delay of any flow of the state of the art that take into account serialization constraints on frame transmissions.

The performance of existing bounds is illustrated on a set of representative automotive examples.

The different protocols imposed particular constraints in the wireless sensor networks. One of these constraints stems from the limitation of the energy source: the sensors are usually powered by a battery. The most important point of consumption is the sending of data. Optimizing communication protocols to minimize consumption is crucial. In this discussion, we propose an original study using the representation of a network of wireless sensors by a grayscale image to construct routing protocols minimizing energy consumption.

Data sharing in vehicular *ad hoc* networks is a challenging task, since the velocity of vehicles and the propagation environment have a substantial influence on the network topology. Three main aspects are considered in this approach: diffusion, routing and security. The first one defines how data are diffused to a destination or some destinations. The second one determines which path among the network is used to support data from a source to a destination. The last one refers to security issues in vehicular networks. In order to highlight issues related to diffusion, routing and security, an introduction to vehicular networks is also given, introducing important terms and concepts and detailing current challenges.

Finally, we introduce different new hybrid topologies such as the new graphiton models described here as trivalent graphs which encode topologically binary information. They allow defining intrinsic discrete spaces which constitute supernode crystals. Besides encoding its own metric, the model supports disturbances due to fault tolerance through the redundancy of information in the paths of connection between supernodes. This new model is expected to resolve the real-time problems encountered in the vehicular network such as routing and security.

Then, we discuss wireless sensor networks (WSN) based on the IEEE 802.15.4 standard. We present, in this discussion, a new topology construction approach. The presented approach is based on the exploitation of RF front-end capabilities in treating multipath signals, thus avoiding the introduction of beacon or superframes scheduling algorithms. Avoiding the introduction of scheduling algorithms ensures a simple solution that could be easily implemented and executed by sensor nodes.

Finally, we introduce the new topology of networks based on the pancake graphs as an alternative to the hypercube for interconnecting processors in parallel computer networks. This network offers attractive and desirable properties. We present, in this discussion, the many-to-one embedding of multiply-twisted hypercube into the pancake networks with dilation 5. These new concepts of this topology are not yet exploited in the vehicular *ad hoc* networks, and this is another future challenge.

Smain FEMMAM
June 2017

Introduction

Recent advances in the integration of communication and sensor technologies have triggered the deployment of numerous attractive applications for road transportation systems. In this regard, the network of connected vehicles contributes to the main building block of intelligent transportation systems (ITS), and the basis for a diversity of applications that can enhance the safety and comfort of road transportation such as accident prevention and road traffic condition.

In order to have a reliable system and good real-time performance, underlying technologies such as real-time communications, routing and security must be reliable and real-time in the first place. On the one hand, secure communication is a key element in wireless networks in order to provide Quality of Service (QoS) for mobile users. On the other hand, robust topologies are needed to ensure efficient cooperation of routing and data diffusion in vehicular *ad hoc* networks.

The proposed new approaches and their efficiency to resolve the problems encountered in the wireless networks are discussed in this book.

We start with a survey of switched Ethernet solutions for real-time audio/video communications in Chapter 1. Switched Ethernet technology is a well-established solution for Local Area Network (LAN) communications.

Introduction written by Smain FEMMAM.

It has more recently been considered as a solution for the network infrastructure on board of transportation systems. It first came with the need to update the firmware of electronic control units (ECUs) due to the development of advanced functionalities for vehicles (ADAS, Power train, Chassis, Driver assistance, etc.), increasing the size of the firmware and the time required to flash it. The introduction of Ethernet technology was then considered for more constrained contexts, e.g. for audio/video (entertainment) and control command real-time (RT) communications.

Chapter 1 is a survey of these evolutions with a particular focus on audio/video communications. In this chapter, we first review existing Switch Ethernet solutions in the context of Automotive, Avionic and Industrial domains.

We focus on the current solutions for audio/video communications, i.e. Ethernet audio video bridging (AVB) specification. This specification has adapted Ethernet technology to meet electromagnetic constraints (EMC) specific to the automotive domain, and at the same time reducing the number of wires required to communicate (only one twisted pair in the current specification for at a throughput of 1 Gb/s instead of four). We review Physical (Broad R-Reach PHY) and Ethernet MAC layer in that perspective. Several new protocols including: clock synchronization (802.1AS gPTP), stream reservation and transmission protocols (IEEE 802.1Qat SRP and IEEE 1722) have been developed, and new mechanisms for forwarding and queuing frames have been proposed.

Then, we study how Ethernet AVB can be used to support RT communications. Ethernet AVB introduces new classes (A and B) with the class-based shaping algorithm (CBSA) for arbitrating between classes A, B and best-effort (with no CBSA) flows. We explain the principles of the CBSA, and we show that it is also benefit to best-effort flows.

Finally, we study how to characterize the worst-case end-to-end delay of any class A, B and best-effort flow. We review two existing approaches: the holistic and the trajectory approaches. The holistic approach considers, for a flow, a safe worst-case arrival scenario on each visited node (possibly non-possible). The trajectory only considers possible scenarios. We characterize the worst-case end-to-end delays obtained with the two approaches for a representative automotive configuration, showing that the trajectory clearly outperforms the holistic approach. We discuss Ethernet AVB's compatibility with most worst-case end-to-end delay requirements, and particularly for audio/video communications. In the following paragraphs, we discuss the application of switched Ethernet-based real-time solutions in different industrial domains including audio/video communications.

Chapter 2 discusses the representation of networks of wireless sensors with a grayscale image. Applications to routing are introduced. Wireless sensor networks are composed of nodes with wireless communication capacity. Usually, these nodes are covering some space they have to monitor. They can be deployed in a hostile environment without easy access. In this case, the power they require to operate are provided by batteries. The energy capacity of this battery is directly related to the lifetime of the sensors. Many research topics are focused on optimizing the energy consumption of the nodes. This energy consumption is due to data sensing, processing and transmission. As wireless communication is the main consumer item, researchers will try to minimize transmissions.

Communications in wireless sensor networks consist of forwarding packets from a sensor to a monitoring station through an *ad hoc* network. In this process, routing protocols are crucial and their strategies dramatically influence the energy consumption of the network. Energy routing protocols are proposed to respond to this challenge.

Our approach introduces a new way to elaborate energy routing protocols based on grayscale images. We use a grayscale image analogy of the wireless sensor network, where each sensor is a pixel and its gray level represents its battery capacity. By convention, a node with high energy capacity will be represented by a bright pixel. The routing process consists then of constructing a path in the bright part of the image. To identify these bright regions, image processing algorithms are used. A convolution filter, such as the Mean, Gauss or Sobel filter, is involved in constructing routing algorithms. This chapter is concluded wtih some simulations on these routing protocols. In this chapter, a new method using a gray level image analogy of wireless sensor networks is introduced. A gray level image is constructed, where each pixel represents a sensor and its gray level represents the battery capacity of this sensor. On this representation of the network, different image processing algorithms are used to produce routing algorithms for the network.

Chapter 3 is dealing with the routing and data diffusion in vehicular *ad hoc* networks.

Data delivery is a crucial task in vehicular networks since current applications require the cooperation of each and every vehicle. Current autonomous vehicles rely on embedded sensors to obtain a local vision of the environment to take decisions. However, the construction of a global vision requires the exchange of information through network communication. In such a context, the IEEE 802.11p provides a communication standard to construct a network dedicated to vehicles. Applications' performance depending on the data delivery service must be reliable, despite the high dynamic of the network topology. This service is investigated through three aspects: first, how to select the destination; second, how to route data; and finally, how to secure communication. The first two are supported by routing protocols, ensuring end-to-end communication and routing data according to the delivery scheme. Suitable in wireless networks, link quality estimators have been widely used by routing protocols dedicated to VANET. Their design must take into account the volatility of the link quality often ranked as reliable, unreliable and bad. However, any estimators meet all the requirements and let this topic still opened. Regarding routing protocols, several strategies have been deployed both to reduce routing messages and get the maximum packet delivery ratio and minimize the end-to-end delay. There is no silver bullet and many strategies have been shown as the best, since they have been designed for a specific situation and take into account only some network features. Moreover, security aspects should be taken into account based on the following requirements: authenticity, integrity, non-repudiation, availability, access control and confidentiality. All these requirements ensure VANET security avoiding attacks on network performance, privacy data and traceability, keeping in mind that VANET prevent risk injury. Two kinds of security must be considered, passive attacks and active attacks. Within passive attacks, only monitoring tasks are performed, unlike in active attacks wherein an action is performed by a hacker. This chapter is organized as follows: first, we describe the context and challenges related to each considered aspect of the data delivery service. Then, we discuss the routing protocols related to vehicular networks. Finally, we introduce the security aspects.

In Chapter 4, we present another hybrid topology based on a new "Graphiton" Model: a computational discrete space, self-encoded as a trivalent graph. The new graphiton models described here are trivalent graphs which encode topologically binary information. They permit defining intrinsic discrete spaces which constitute supernode crystals. Besides encoding its own metric, the model supports disturbances due to fault tolerance through the redundancy of information in the paths of connection between supernodes. Coming from theoretical physics, they may find applications in network management and artificial intelligence. For the first time, an information system structure, rich enough to model the universe itself, but relying ultimately on set theory, traverses set theory, topology, information theory, graph theory, geometry, algebra, theoretical physics and even computer and wireless network science in a logical, straightforward and elegant way.

Chapter 5 presents a beacon cluster-tree construction approach for ZigBee/IEEE802.15.4 networks. Wireless sensor networks (WSN) based on the IEEE 802.15.4 standard are constantly expanding. Applications like production control and building control are increasingly based on WSN because of their energy efficiency, self-organizing capacity and protocol flexibility. The IEEE 802.15.4 standard defines three network topologies: the mesh topology, the star topology and the cluster-tree topology. However, the construction of cluster-tree networks based on the beacon mode is still undefined by the IEEE 802.15.4 standard. A beacon cluster-tree topology has the advantage of giving all the benefits of the beacon mode (Synchronization, QoS support through Guaranteed Time Slots) and, at the same time, allows the construction of large networks to cover large areas. In order to enable the construction of such topology, i.e. beacon cluster-tree, we present, in this chapter, a new topology construction approach. The presented approach is based on the exploitation of RF front-end capabilities in treating multipath signals and, thus, avoiding the introduction of beacon or superframes scheduling algorithms. Avoiding the introduction of scheduling algorithms ensures a simple solution that could be easily implemented and executed by sensor nodes in wireless networks. Finally, a geo-location application is detailed.

In Chapter 6, we propose a new hybrid topology based on embedding the twisted hypercube into pancake graphs. Among Cayley graphs on the symmetric group, the pancake graph is a viable interconnection scheme for parallel computers, which has been examined by a number of researchers. The pancake was proposed as an alternative to the hypercube for interconnecting processors in parallel computers. Some attractive properties of this interconnection network include: vertex symmetry, small degree, a sub-logarithmic diameter, extendibility, high connectivity (robustness), easy routing, regularity of topology, fault tolerance, extensibility and embeddability of other topologies. In this chapter, we present the many-to-one dilation 5 embedding of an n-dimensional twisted hypercube into an n-dimensional pancake. This new topology provides a reliable system and good real-time performance in construction of large networks. In our knowledge, this topology has not been used until now in real-time embedded systems, particularly in vehicular *ad hoc* networks. We hope that this chapter will help researchers and practitioners develop an interest in this area of application.

1

A Survey of Switched Ethernet Solutions for Real-time Audio/Video Communications

1.1. Introduction

Ethernet is attractive to a number of industrial domains, like the automotive and aviation sectors as well as the industrial control domain, due to the fact that it provides a good bandwidth with mass-market products that make it a low-cost solution. However, early models of Ethernet, such as Ethernet with CSMA/CD [IEE 00] (carrier-sense multiple access with collision detection), introduce collisions and therefore cannot guarantee real-time performances, like deterministic delay, low jitter, etc. Switched Ethernet-based solutions eliminate collisions by employing full-duplex communication links. Although jitters can be introduced at switch output port buffers, several approaches have considered deterministic delay analyses that guarantee the real-time performance of such networks. Especially with new emergent technologies, like AVB, traffic shaping is employed in order to fulfill varying real-time requirements imposed by different data types and different flow classes. In the following sections, we discuss the application of switched Ethernet-based real-time solutions in different industrial domains including audio/video communications.

Chapter written by Xiaoting LI and Laurent GEORGE.

1.1.1. *Automotive industry*

Automotive electronic systems are experiencing rapid growth thanks to the development of advanced driver-assistance systems (ADAS). On the other hand, original equipment manufacturers (OEMs) require more and more functions and applications for different in-vehicle domains in order to enhance driving safety as well as passengers' comfort. For example, infotainment systems provide both information (navigation system, remote diagnostic, etc.) and entertainment (DVD play) services. All these new applications imply a greatly increased data exchange with heterogeneous data types, including control messages, camera streaming and audio/video streaming among others.

The applications are implemented in electronic control units (ECUs), which were first connected by point-to-point links and later replaced by multiplexed communications based on embedded networks. The most commonly deployed in-vehicle embedded networks are CAN (controller area network) and/or FlexRay as back-bone and control networks, LIN (local interconnect network) as a low-cost broadcast master-slave serial communication bus used for connecting less important in-vehicle elements, and MOST (media-oriented systems transport) as a media signals transmission solution based on optical fiber cables and ring topology with a high throughput of 150 Mbps for MOST150. Other non-automotive-specific standards have also been proposed for camera-based ADAS applications. LVDS (low-voltage differential signaling) offers a high bandwidth of up to 655 Mbps based on twisted pair copper cables, which makes it a very attractive candidate for automotive camera manufacturers. IEEE 1394, also known as Firewire, is an interface standard defined as a serial bus for high-speed communications and isochronous real-time data transfer. It is often used in consumer video cameras and was also considered as a candidate for automotive infotainment systems.

The currently available tools cannot fulfill the new requirements of automotive electronic systems with increased amounts of data exchange in terms of bandwidth as well as increased system complexity due to the different communication protocols involved. Moreover, current solutions are often automotive-specific, which means that the development to upgrade them is expensive. Therefore, OMEs are looking for new low-cost technologies. Switched Ethernet-based solutions offer a good bandwidth and are standardized with mass-market technological components which allows OMEs to reduce the development cost, making these eligible and attractive candidates for the automotive domain.

Several Ethernet-based solutions have been presented and discussed for automotive applications, including IEEE 802.1Q [IEE 11c], AVB switched Ethernet and Time-Triggered Ethernet (TTEthernet) [STE 08]. IEEE 802.1Q enables priorities associated with frames by inserting a VLAN (Virtual LAN) tag in the Ethernet header, between the source MAC address and the TLV (Type/Length Value). This solution was originally intended for dividing a physical LAN into several logical LANs in order to limit the broadcast traffic. Later, this technique was introduced in the automotive sector as shown in [LIM 12]. In VLAN tags, there is a 3-bit dedicated field for storing priority values which allows for up to eight different priorities. TTEthernet is used in *both* the automotive and aviation industries and will be presented in section 1.1.2.

IEEE Ethernet audio/video bridging (AVB) [IEE 11a, IEE 09, IEE 10, IEE 11b] switching shows great potential as it provides dedicated credit-based bandwidth for real-time traffic which relies on a credit-based shaping algorithm (CBSA). Furthermore, thanks to the CBSA, the starvation of non-real-time traffic with lower priority can also be avoided. The major challenge for the adoption of this AVB switching technology in the automotive industry is the possibility of encountering hard real-time constraints. The timing property of real-time traffic should be predictable (hence proved correct by construction), which leads us to the problem of characterizing the worst-case end-to-end delay of any flow transmitted in the network. The details of the real-time performance analysis of AVB switching will be discussed later.

1.1.2. *Aviation industry*

Avionic electronic systems have experienced similar evaluations as automotive electronic systems. At the beginning, each application was implemented in an LRU (line-replaceable unit) and LRUs were connected by point-to-point links. One of the first standards specifically targeted at civil avionic applications is ARINC 429, which defines a multi-drop field bus link connecting one transmitter and several receivers. With increased avionic applications, the implementation of a multi-drop bus for each transmitter makes the interlinking dense and therefore leads to an increased weight and a complex system design. Boeing and NASA jointly developed a new technology, ARINC 629, based on a triplex-bus layout, including dual redundancy buses and a flight control bus. It allows for multi-access by employing a CSMA/CD media access mode and avoids collisions by using a time-division-based mechanism. However, the ARINC 629 technology is an expensive solution in terms of development cost due to its high complexity and it is almost exclusively implemented in Airline Boeing 777.

For the purpose of reducing development costs, Avionic Full DupleX (AFDX) switched Ethernet, based on COTS (commercial off-the-shelf) technology, was proposed. AFDX is defined by ARINC 664 part 7 and has been successfully integrated into the Super Jumbo Airbus A380.

An AFDX switched Ethernet network is composed of end-systems interconnected by a switched Ethernet network via full-duplex links. It offers a typical bandwidth of 100 Mbps. An end-system provides network access with traffic shaping policies to avionics applications and functions hosted on processing modules. An AFDX switch forwards frames towards the destination end-systems based on a statically defined forwarding table. Each switch uses store-and-forward mode in order to check the frame integrity and reject the invalid frames. Each communication over an AFDX network is defined by a VL (Virtual Link), which defines a logical unidirectional connection from one source end-system to one or several destination end-systems. This multicast characteristic represents an analogy to the ARINC 429 multi-drop bus thanks to the VL concept. Each VL is associated with limited bandwidth utilization and is characterized by a limited rate, called BAG (bandwidth allocation gap), and a limited frame length. The attributed quota of each VL is ensured at each emitting end-system by a traffic shaping technique and then checked at each switch. Frames not respecting the VL constraint are rejected during transmission. For safety considerations, two redundant separate AFDX networks, network A and network B, are both implemented. Each transmitting end-system sends the same frame to two separate redundant networks and each receiving end-system integrates redundancy management to make sure that the first arrival valid frame is accepted. Thanks to the VL concept, with its traffic shaping technique, static configuration and redundant networks, AFDX supports avionics applications with real-time requirements and enables the deterministic behavior of the network.

TTEthernet (Time-Triggered Ethernet) is another Ethernet-based solution for real-time applications. It was developed and marketed by TTTech for safety-related applications and later standardized by SAE International as SAE AS6802. It has been adopted by Honeywell for production programs in the aerospace and automation industries.

TTEthernet integrates both time-triggered and event-triggered services over Ethernet. The objective of designing such networks is to provide a seamless communication to all applications, including safety-critical applications from closed industrial networks as well as conventional applications (e.g. web service). This solution allows three different types of traffic: time-triggered (TT) traffic, rate-constrained (RC) traffic and best-effort (BE) traffic to be transmitted through the same network infrastructure. The system-wide synchrony is established for TT service by using a unique synchronization mechanism that can be combined with other mechanisms such as IEEE 1588. TT traffic is then exchanged without collisions at predefined times with a precision at the single microsecond level, which makes the TT service deterministic and eligible for real-time high-critical applications. An RC service does not need system-wide synchronization, since it works based on a predefined bandwidth associated with each RC flow thanks to a traffic shaping technique. The real-time performance of RC traffic is also guaranteed although jitter can be introduced at the buffer level, and thus RC traffic can be applied to less stringent deterministic distributed systems (automotive and aerospace) or also multimedia systems. Finally, BE traffic uses the remaining bandwidth and no QoS is guaranteed.

1.1.3. *Industrial automation*

Field buses are also being replaced by Ethernet in the industrial automation domain. Current technologies include PROFINET, EtherNet/IP, CC-Link IE, Secors III, Powerlink, Modbus/TCP and EtherCAT. Communication in the domain of industrial automation is somewhat based on Master/Slave mode where one device or process has unidirectional control over one or several other devices. Different principles and approaches have been developed depending on the real-time and cost requirements.

Three different classes, class A, class B and class C, are defined by considering the implementation of slave device. Class A uses unmodified standard Ethernet hardware and TCP/IP protocol suite. However, the real-time performance cannot be guaranteed because of the TCP/IP stack. Class B still uses unmodified standard Ethernet hardware, but enables a dedicated layer directly connected to the MAC layer in order to bypass the TCP/IP stack. Therefore, class B provides soft real-time performance. Class C employs dedicated hardware in order to provide higher real-time performance guarantees without excluding the use of TCP/IP. In the following sections, several existing industrial automation Ethernet technologies are reviewed.

PROFINET International maintains PROFINET, an industry technical standard developed for data communication over industrial Ethernet. PROFINET supports line and tree topologies. It has three versions corresponding to the three classes. PROFINETv1.2, also called PROFINET TCP/IP, is a class A protocol based on TCP/IP stack. It allows access to existing PROFIBUS networks via Proxy devices for parameter data. Nowadays, this technology is no longer supported by PROFINET International. PROFINETv2, also called PROFINET RT (Real-Time), is a class B protocol bypassing TCP/IP stack for real-time process data, and therefore enhances a soft real-time performance and is considered an alternative of PROFIBUS. PROFINETv3, also called PROFINET IRT (Isochronous Real-Time), is a class C protocol with an implementation of a dedicated hardware. Hard real-time performance is achieved of PROFINET IRT by employing time slicing mechanism, enabling a bandwidth reservation of real-time data like motion control applications.

Powerlink is an open real-time and deterministic industrial protocol based on standard Ethernet. It uses hubs with half duplex communication based on the access mode CDMA/CD in order to reduce delays. For the purpose of deterministic communication, it integrates a polling method with time slicing mechanisms. All frames are broadcasted through the network. The master broadcasts a request to all slaves and waits for the response from a slave before broadcasting another request. Line topology is supported, but the number of nodes connected to a hub is limited according to the performance consideration.

Modbus is a widely accepted serial communication protocol thanks to its simplicity. It was expended to Modbus TCP based on the TCP/IP stack, and therefore, it is a class A protocol, which means that it cannot guarantee real-time performance. It allows several transactions which means that the master (client) does not need to wait for the response from a slave (server) before starting another request.

1.2. Ethernet AVB solution

Initially proposed as a time-synchronized solution for audio and video transmission of live environments, audio video bridging (AVB) switched Ethernet has been quickly considered for use in the automotive sector due to its capacity to guarantee real-time communication with a bandwidth of 100 Mbps. The IEEE 802.1 working group, audio/video bridging task group, developed a comprehensive set of AVB specifications for the purpose of providing a reliable and low latency service for synchronized audio video streaming.

1.2.1. *Physical and MAC layers*

AVB is a solution based on Ethernet (IEEE 802.3) for transmission of audio and video streaming. It involves layers 1 and 2 of the OSI (Open Systems Interconnection) reference model, that is the physical layer and the media access control (MAC) layer.

Instead of using the classical Ethernet unshielded twisted pair (UTP) cable with four pairs, automotive manufacturers propose the use of unshielded twisted single-pair cabling, which allows all vehicle components to connect using lighter and more cost-effective wires, therefore saving weight and space of in-vehicle networks. New technologies have been proposed to enable simultaneous transmit and receive (i.e. full-duplex) operations on a single-pair cable. For example, BroadR-Reach automotive Ethernet standard, developed by a network equipment vendor Broadcom, realizes a full-duplex communication based on single-pair cables with a bandwidth reaching 100 Mbps. This standard (also known as 100BASE-T1) was later standardized by IEEE 802.3 in IEEE 802.3bw-2015 Clause 96. It draws great interest from several interest groups including the OPEN Alliance SIG, which is formed by several automotive and equipment companies, showing great potential to be widely adopted in automotive Ethernet as the connectivity standard.

However, as indicated in [TUO 15], the bandwidth of 100 Mbps guaranteed by BroadR-Reach can only carry compressed video streams. The need for an

Ethernet with higher bandwidth has been addressed by IEEE Reduced Twisted Pair Gigabit Ethernet (RTPGE) Study Group. The aim of the study is to modify the current IEEE 802.3 standard in order to support 1 Gbps Ethernet on fewer than three twisted copper pairs and therefore to expend the use of Ethernet in automotive as a communication network.

The AVB specifications consist of the following IEEE standards:

– IEEE 802.1AS – Timing and Synchronization for Time-Sensitive Applications;

– IEEE 802.1Qat – Stream Reservation Protocol (SRP);

– IEEE 802.1Qav – Forwarding and Queuing Enhancements for Time-Sensitive Streams;

– IEEE 802.1BA – Audio Video Bridging Systems.

Some important issues of the AVB standard set are discussed in the following sections.

1.2.2. *Ethernet AVB standard family*

1.2.2.1. *IEEE 802.1AS gPTP*

IEEE 802.1AS is a synchronization standard based on IEEE 1588. IEEE 1588 standardized the use of physical layer timestamps in order to reach a time accuracy at sub-microsecond level. Timestamp, based on a synchronization physical layer event: the start of frame, is used to measure and compute network delays.

IEEE 802.1AS is a specific profile of IEEE 1588 with fewer options and extended physical layer options. The time distribution of IEEE 802.1AS is processed based on a hierarchical clock architecture where a master clock distributes the "precise origin timestamp" periodically, e.g. every 100 ms, by two possible ways. The first one is called two-step messaging that a synchronization frame Sync is sent first and then followed by a FollowUp frame used to contain the timestamp of the Sync frame. An illustration of two-step messaging is given in Figure 1.1. At time instant $t1$, the Sync frame is sent from the master clock and its departure time $t1$ is sent within the frame FollowUp. The second way requires only one step that demands the physical layer to insert the timestamp into the transmitting Sync frame "on the fly". In this case, the FollowUp frame is no longer needed.

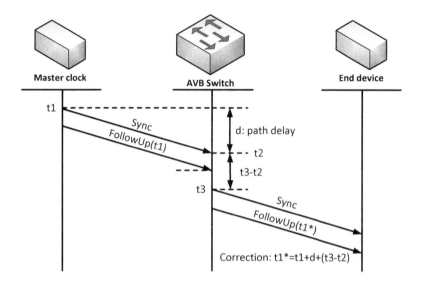

Figure 1.1. *An illustration of clock synchronization*

Each non-master bridge forwards the Sync frame as well as the FollowUp frame (if any) with a correction field to the following network equipment till the end device which is a slave in the synchronization process. This type of bridge is called transparent clock. The correction field is used to contain information about accumulated upstream path delay and switching latency consumed by a bridge to forward the Sync frame to the output port. As illustrated in Figure 1.1, frame Sync is received at the AVB switch, a transport clock, at time instant $t2$ after experiencing a path delay d on the link between the master clock and the AVB switch. The frame Sync is then forwarded to the end device at time instant $t3$. The AVB switch computes the correction by adding the path delay d and the switching latency $(t3 - t2)$ and then send it in the correction field of the FollowUP frame as $t1^*$ shown in Figure 1.1. At each end device, the corrected time is computed by the sum of origin timestamp distributed by the master clock, the correction field and the upstream path delay. Each pair of bridges measure the path delay of the Ethernet connection between them.

In order to obtain a better guarantee on the accuracy of the clock synchronization, clock management frames (Sync, FollowUP, etc.) can be sent in a higher priority VLAN based on IEEE 802.1Q.

1.2.2.2. *IEEE802.1Qat SRP*

IEEE 802.1Qat SRP provides an on-line stream reservation mechanism based on Multiple Stream Registration Protocol (MSRP). A Talker, an end station that is the source of a stream, sends a SRP request (also called a declaration) to a Listener, a destination end station that is a receiver of the stream. Each bridge along the path propagates the request and reserves the required resource, e.g. the bandwidth. If the resource along the path can guarantee the transmission of the stream with required Quality of Service, the Listener which is requesting attachment to the stream confirm the request by sending back a SRP response declaration indicating "Listener Ready". Otherwise, it sends a "Listener Asking Failed" response declaration to the Talker. Note that one and more Listeners can attach to the same stream, which indicates a multi-cast communication.

IEEE 802.1Qat SRP is a dynamic resource allocation protocol. Once a stream is not any more needed or produced, a Listener or a Talker can issue an SRP message to withdraw the required resource declaration and therefore the resource can be released and re-allocated to other streams on demand.

1.2.2.3. *IEEE 1722*

IEEE 1722 is a Layer 2 AVB transport protocol, also called AVTP. It enables interoperable streaming through bridges by defining media formats and encapsulations, media synchronization mechanisms as well as stream Multicast Address Allocation Protocol (MAAP). It requires that all devices sending, receiving or forwarding stream data support services provided by IEEE 802.1AS for synchronization as well as IEEE 802.1Qat SRP and 802.1Qav for Quality of Service (QoS).

IEEE 1722 supports raw and compressed audio/video data formats of applications using IEC 61883. Each stream has a unique ID and can be streamed from a Talker to one or several Listeners based on a multicast address allocated by the MAAP. Listeners can be rendering devices like multi-display (compound) screens.

Each stream data is encapsulated in a standard Ethernet Layer 2 frame with an inserted presentation time. The presentation time is the time at which this audio/video stream data should be rendered on the destination devices. It is the Talker who is responsible for setting the presentation time by adding the worst-case transmission delay of the data to the sampled time according to the IEEE 802.1AS clock. Based on a synchronized presentation time, all the rendering devices play the stream at the same time.

IEEE 1722 also supports time-sensitive control stream from field buses, e.g. CAN, FlexRay and LIN, which are connected to AVB Ethernet through a gateway.

1.2.3. *Evolution: Ethernet TSN*

In the case of highly time-constrained flows, a flow even with the highest priority can suffer from a non-preemptive delay in each visited switch. This non-preemptive delay corresponds to the case where a low priority frame has just started to be sent before the arrival of the high priority frame. As frame transmission is non-preemptive, the high priority frame has to wait for the end of the low priority frame to be transmitted. An example of applications that require deterministic communications with small jitters in the context of automotive applications is control applications, where data sent is used to feed the parameters of control loops that are critical for the safety of the car. Those frames carrying control data are transmitted on a repeating time schedule. A late delivery of such frames can lead to instability, inaccuracy or failure of the operation of the control loops concerned. In the context of autonomous vehicles, trajectory control of a car assisted by cameras will become more and more important. Assuring a predictable transmission of the video captured by the cameras can become critical, e.g. for obstacle detection or abnormal situations detection.

Time-Sensitive Networking (TSN) [THI 16] is a set of standards under development by the Time-Sensitive Networking task group of the IEEE 802.1 working group. The goal of Ethernet TSN is to provide a solution for the transmission of time-critical control traffic with other classes of traffic in the same network, as long as such mixing can be achieved while still meeting the timing requirements of the time-critical traffic.

The time critical traffic is a scheduled traffic sent in protected windows of fixed sizes. A periodic pattern of protected windows must be defined on each switch such that any time critical flow can be sure to be blocked only by other time critical traffic. A guard-band of duration equal to the transmission of a frame with the maximum payload is set before any protected window. A guard-band can only be used by non-critical frames. A non-time critical frame cannot start its transmission in a guard band but can only complete its transmission. This potentially leads to impose idle times in the transmission of frames to preserve scheduled windows. Figure 1.2 shows an example of two periodic patterns set on two switches SW1 and SW2 with corresponding protected windows.

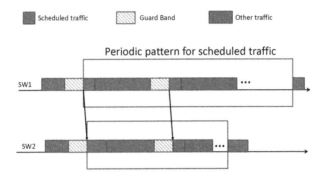

Figure 1.2. *Scheduled traffic principles*

1.3. AVB deterministic RT communications

Several Ethernet-based solutions have been presented and discussed for industrial applications, including Ethernet 802.1Q, Audio Video Bridging switched Ethernet, Time-Triggered Ethernet as well as Avionics Full-DupleX (AFDX) switched Ethernet. Among them, the IEEE AVB switching shows great interest as it provides a dedicated credit-based bandwidth for real-time traffic relying on a credit-based shaping algorithm (CBSA). This feature is especially attractive to the automotive domain to deal with the growing demand of driver assistance and infotainment systems. Furthermore, thanks to the CBSA, the starvation of non-real-time traffic with lower priority can also be avoided. The major challenge for the adoption of the AVB switching technology in the industrial control domain is the possibility to meet hard real-time constraints. The timing property of real-time traffics should be predictable (hence proved correct by construction), and therefore, the worst-case end-to-end delay of any flow transmitted in an AVB switching network must be characterized.

1.3.1. *Class A/B RT constraints*

The AVB standard 802.1Qav supports eight traffic classes, including at least one traffic class that can support traffic which is not subject to bandwidth reservation, such as "best-effort" traffic. Without loss of generality, three traffic classes with different priorities are considered, including two AVB stream reservation (SR) classes, denoted class A and class B, with credit-based shaping algorithm (CBSA) as well as one non-SR class, denoted class C, for soft real-time "best-effort" flows without CBSA. According to the real-time requirements of AVB standard, class A flows should have a maximum delay of 2 ms and class B flows should have a maximum delay of 50 ms for seven hops. We consider the possibility to support soft real-time flows in a class C "best-effort" class. This corresponds to non-AVB compliant end-systems sending their flows with 802.1Q protocol (with no CBSA). It can be interesting for such flows to determine whether they can support soft real-time constraints, taking into account for such flows the impact of higher priority class A and class B AVB flows with CBSA. The flows of the same class (A, B or C) can be served in first-in first-out (FIFO) order.

1.3.2. *CBSA (credit-based shaping algorithm)*

AVB switching integrates a class-based non-preemptive scheduling of flows in all AVB output ports. All incoming flows are forwarded to a specific output port queue, according to their class defined by a static priority. AVB SR class A has a higher fixed priority than AVB SR class B. Flows of an AVB SR class are buffered in a corresponding FIFO queue at each output port and their traffic is shaped by an AVB traffic shaper.

Except AVB SR classes, the "best-effort" non-SR class C flows have the lowest priority in the network and they are put in a FIFO queue at each output port without traffic shaping.

Each AVB supporting output port, denoted h, imposes a CBSA for each AVB SR class. Flows in each SR class X ($X \in \{A, B\}$) are associated with a credit-based mechanism, called *SendSlope* (in bits per second), which decreases at a rate $-\alpha_X^-(h)$ at an output port h when transmitting class X frames, until the credit drains (below zero). Then, the credit replenishes at a rate $\alpha_X^+(h)$ (in bits per second), called *IdleSlope*, until the credit becomes higher than or equal to zero. During the credit replenishing time, class X frames are pending in the buffer of class X. The transmission rate of a link connected to an AVB supporting port is denoted R. According to the AVB standard 802.1Qav 8.6.8.2, the values of *SendSlope* and *IdleSlope* are defined as follows:

$$\alpha_X^+(h) \leq R$$

$$-\alpha_X^-(h) = \alpha_X^+(h) - R$$

As a result, a bandwidth fraction of the link connected to the output port h reserved by class X can be computed by:

$$\frac{\alpha_X^+(h)}{\alpha_X^+(h) + \alpha_X^-(h)}$$

The transmissions of class X frames cannot use more bandwidth than the reserved fraction. It means that class X frames can be transmitted only when the credit of class X is non-negative.

The credit of an AVB SR class, if positive, is immediately set to zero after the transmission of the last pending frame in the corresponding class buffer.

The "best-effort" class C flows do not experience an AVB CBSA at a switch output port. However, since they have the lowest priority in the network, they can be transmitted only when (1) no frame of any AVB SR class is pending; or (2) the credits of all AVB SR classes drain.

An example of AVB SR class credit is shown in Figure 1.3. There are two SR classes A and B as well as a non-SR class C transmitted at an AVB supporting switch output port h.

Frames A_1, A_2 and A_3 are class A frames. They arrive at the output port h at the same time and they are put in the class A buffer in the order indicated in Figure 1.3. Frames B_1 and B_2 are class B frames arriving at the output port h at the same time as class A frames and put in the class B buffer as shown in Figure 1.3. Frames C_1 and C_2 are class C frames arriving at the class C buffer at the same time. If frame C_1 arrives just before frames A_1 and B_1, it will be sent first and will therefore delay the transmission of frame A_1 as well as the transmission of frame B_1 due to non-preemptive frame transmission.

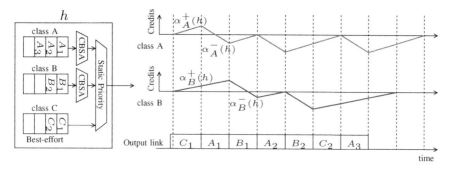

Figure 1.3. *Illustration of an AVB supporting switch output port*

The credit of class A starts to accumulate at a rate $\alpha_A^+(h)$ until the end of the transmission of frame C_1 and then starts to decrease at the rate $-\alpha_A^-(h)$ as soon as frame A_1 starts to be transmitted. Meanwhile, the credit of class B accumulates at a rate $\alpha_B^+(h)$ as frame B_1 is pending during the transmissions of frame C_1 (non-preemptive delay) and of frame A_1 (higher priority). However, after the transmission of frame A_1, frame B_1 starts to be transmitted even if there are still class A frames waiting in the buffer since the credit of class A is negative whereas the credit of class B is positive. During the transmission of frame B_1, the credit of class B decreases at a rate $-\alpha_B^-(h)$, while the credit of class A starts to replenish and reaches zero at the end of the transmission of frame B_1. This enables class A frames to be eligible for transmission just after frame B_1 transmission. This process continues until all the waiting frames at the output port h are transmitted.

One characteristic that distinguishes the AVB switching from a static priority scheduling is that a frame in one class can only be transmitted if the credit of its class is higher than or equal to 0. If the credit drains, pending frames cannot be transmitted even if they have the highest fixed priority. In that case, lower priority frames can be transmitted first. This characteristic has been illustrated by frame B_1 being transmitted before frame A_2 due to negative credit of class A when B_1 started its transmission (see Figure 1.3). It is also illustrated by the transmission of frame C_2, a "best-effort" frame, when the credits of both class A and class B drain.

According to the AVB standard 802.1Qav 34.3.2, the *IdleSlope* of an AVB SR class X can be set by two ways:

– If the stream reservation mechanisms defined in Stream Reservation Protocol (SRP) are supported, then the value of $\alpha_X^+(h)$ is obtained by the SRP.

– Otherwise, the value of $\alpha_X^+(h)$ can be set by a management functionality of the switch.

1.3.3. *End-to-end communication delays*

The end-to-end communication delay of a frame of a flow is the temporal duration between the instant when the frame arrives at the output port buffer of its source end-system and the instant when the frame is received by the destination end-system. It consists of several parts, including the transmission delay of the frame on each link along the path, the waiting delay caused by other interference frames at each output port buffer, the switching latency at each visited switch used to deal with the forwarding as well as the delays imposed by a CBSA shaper when the credit is negative.

Several works have been devoted to delay analysis on an AVB switched Ethernet network. Two approaches can be found in the literature: the simulation approach and the worst-case delay analysis approach. The simulation approach provides valuable distributions of end-to-end delays without guaranteeing worst-case delays. It is useful when comparing the average timing performance between the AVB switching and other scheduling policies without stringent timing requirements, hence not for safety-critical applications. A worst-case delay analysis of AVB switching provides delay upper bounds. This work is mainly based on two approaches: the holistic approach and the trajectory approach.

1.3.3.1. *Holistic approach*

Holistic approach considers the worst-case scenario and computes a delay upper bound on each output port. The worst-case end-to-end delay upper bound is the summation of delay upper bounds in all visited output ports of a path. Work about delay analysis based on holistic approaches includes [DIE 12a, IMT 09, DIE 12b, BOR 14]. A performance study has been conducted in [IMT 09] in which both simulation and analytical results are compared on a standard Ethernet-based case study. A theoretical approach based on a holistic approach [TIN 94] has been introduced for the computation of end-to-end delay upper bounds. This approach is a message-based analysis. It means that this analysis ignores flow models, e.g. periodic or sporadic flows, which can generate several frames interfering with the transmission of the frame under analysis. A worst-case delay analysis of AVB switching has been proposed in [DIE 12a, DIE 12b] based on compositional performance analysis. This approach is a flow-based analysis, but the end-to-end delay is computed by the sum of worst-case delays conducted by a local analysis at each hop. A message-based timing analysis which does not take into account sporadic and periodic flow models has been provided in [REI 13]. This approach has been developed based on a modular performance analysis (MPA), which requires a local timing analysis at each single queue. One main inconvenience of the proposed analysis is the lack of theoretical proofs. A schedulability analysis on a single AVB switch has been shown in [BOR 14]. This work has provided formal proofs and formulas on the worst-case response time of a flow on a single switch basis with partial improvements based on the property of AVB shapers. It has focused on the constrained deadline flow models and needs to be extended with a holistic approach for an AVB switched Ethernet network with several switches.

1.3.3.2. *Trajectory approach*

The trajectory approach (TA) was first proposed for first-in first-out (FIFO) scheduling in [MAR 06a] and then extended to non-preemptive fixed priority/first-in first-out (FP/FIFO) scheduling and fixed priority/earliest deadline first (FP/EDF*) scheduling in [MAR 06b, MAR 06c]. This approach has also been applied to the Avionics Full-Duplex Switched Ethernet (AFDX) networks in [BAU 10, BAU 12]. It has shown a great improvement in terms of worst-case end-to-end delay compared with an holistic approach (see [MAR 05, BAU 10]). The main feature distinguishing this approach from holistic approaches is that it considers worst-case end-to-end delay for a flow along its path instead of considering the summation of worst-case delays in each visited node of a path. Therefore, an extended trajectory approach to address worst-case end-to-end delay analysis for sporadic flows sent on an AVB switched Ethernet network was proposed in [LI 16]. In the following, the AVB switched Ethernet network and flows are first introduced and then the trajectory approach extension for AVB switched Ethernet is presented.

1.3.3.2.1. Network and flow models

An AVB switched network is composed of electronic control units (ECUs) interconnected by a full-duplex switched Ethernet network with Audio Video Bridging (AVB) implemented in each switch.

Each ECU sends a set of flows through an output port. At an output port, each traffic class has a buffer supporting a scheduling policy. A traffic shaping technique is adopted at the output port of each ECU in order to guarantee quality of service (QoS) properties by keeping a minimum inter-arrival duration between two consecutive frames of a flow.

Each switch uses a store and forward policy. It receives frames from input ports and forwards them to the corresponding output ports based on a static routing table. At each output port h, each AVB SR class has one buffer supporting AVB traffic shaping, while the "best-effort" flows have one buffer supporting FIFO scheduling. There is a switching latency to deal with frame forwarding between an input port and an output port of a given switch and it is upper bounded by a known value sl.

Links connecting ECUs and switches are full-duplex, which guarantees no collisions on links. All the links of the network transmit frames at the same rate R which is also called link bandwidth. The value of *IdleSlope* $\alpha_X^+(h)$ of each AVB SR class X at the output port h is either decided by the SRP operation if supported or set to a value by a switch management functionality.

There are n sporadic flows transmitted in a network. Each flow τ_i emits a sequence of frames according to its temporal characteristics. A frame emitted by a flow τ_i is denoted f_i, and each frame is an Ethernet standard frame [IEE 12] since AVB switched Ethernet is a standard Ethernet basis solution. A flow τ_i is characterized by a 3-tuple $\{T_i, C_i, Pr_i\}$ defined as follows:

– T_i: the minimum inter-arrival duration between two consecutive frames of flow τ_i;

– C_i: the transmission time of a frame of flow τ_i on a link. For simplicity reasons, the transmission time of a flow τ_i is considered constant for all its frames;

– Pr_i: the fixed priority of flow τ_i.

A periodic flow τ_i is a particular case of a sporadic flow with a constant inter-arrival duration T_i. Each sporadic flow τ_i follows a path statically defined by a sequence of output ports $\mathcal{P}_i = \{first_i, \ldots, last_i\}$, where $first_i$ is the output port of the source ECU of flow τ_i and $last_i$ is the last visited switch output port of flow τ_i along the path $\mathcal{P}_i$.

The index set of flows belonging to a class $X \in \{A, B, C\}$ is denoted $classX$. Since the classes are priority based, two other flow index sets are defined for a flow τ_i ($i \in classX$) as follows:

– hp^X: the index set of flows belonging to a class having a higher priority than the one of class X;

– lp^X: the index set of flows belonging to a class having a lower priority than the one of class X.

An example of a network is depicted in Figure 1.4. It includes eight ECUs denoted ECU_i ($i \in \{1, 2, \ldots, 8\}$), interconnected by three switches, S_1, S_2 and S_3, via full-duplex links. Each ECU has one output port denoted by the same notation as for its ECU. As each switch could have several output ports, the port of a switch is denoted by the switch name followed by the port number. For instance, flow τ_1 is emitted by the output port of its source ECU, i.e. $first_1 = ECU_1$, and then forwarded by the output port 1 of switch S_1, i.e. S_{11}. The transmission continues along the path till its last visited output port 1 of switch S_2, i.e. $last_1 = S_{21}$, and finally, it is received by its destination ECU ECU_8. Hence, flow τ_1 crosses three output ports that compose the path $\mathcal{P}_1 = \{ECU_1, S_{11}, S_{21}\}$.

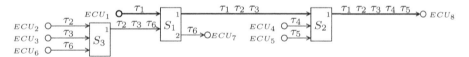

Figure 1.4. *An example of a switched Ethernet network with AVB*

Note that multicast flows are also supported. In the case of a multicast flow, each path is considered independently and the delay analysis is conducted for each path of a multicast flow. The maximum delay of a multicast flow is then the maximum delay among all the paths of the flow. For simplicity reasons, only unicast flows are illustrated in this chapter.

1.3.3.2.2. Trajectory approach extension for AVB

The trajectory approach extension for AVB switched Ethernet is developed based on the network and flow model presented previously. The objective of this approach is to compute the worst-case end-to-end delay upper bound R_i of any class X flow τ_i transmitted in the network.

As already mentioned, one characteristic that distinguishes the AVB switching from a static priority scheduling is that a frame in one class can only be transmitted if the credit of its class is higher than or equal to 0. It means that the CBSA shaping introduces an extra waiting delay when the credit is negative for FIFO scheduling of same class flows. This delay does not impact the frame being transmitted, but it does delay the next frame pending in the buffer in the same class.

An illustration is given in Figure 1.5 where two class X frames f_i and f_j arrive at the same time at the output port h. If they are served by FIFO without CBSA, frame f_i is transmitted first and then delays frame f_j by its transmission time C_i. If they are served by AVB with CBSA, then frame f_i delays frame f_j by its transmission time C_i plus shaping imposed delay due to negative credit of class X. This delay can be upper bounded by:

$$C_i \cdot \frac{\alpha_X^-(h)}{\alpha_X^+(h)}$$

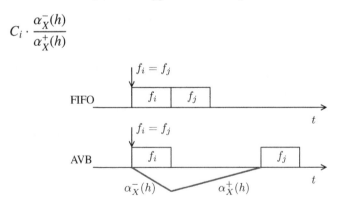

Figure 1.5. *Comparison between FIFO and AVB of class X*

Therefore, the total delay introduced by frame f_i to frame f_j on the output port h is upper bounded by:

$$C_i \cdot \left(1 + \frac{\alpha_X^-(h)}{\alpha_X^+(h)}\right)$$

Since frame f_i and frame f_j can share several output ports along the path and the values of $\alpha_X^-(h)$ ang $\alpha_X^+(h)$ could be variable due to SRP, the maximum delay should be considered by the following computation:

$$C_i \cdot \left(1 + \max_{n \in \mathcal{P}_i \cap \mathcal{P}_j} \left(\frac{\alpha_X^-(h)}{\alpha_X^+(h)}\right)\right) \tag{1.1}$$

Based on the adaptation of CBSA shaping, the worst-case end-to-end delay upper bound of a class X flow τ_i is computed by the equation [1.2]:

$$R_i = \max_{0 \le t \le \mathcal{B}_i} \{W_{i,t}^{last_i} - t + C_i\} \tag{1.2}$$

where $\mathcal{B}_i$ represents the largest possible length of the busy period in each encountered output port of the network, and $W_{i,t}^{last_i}$ is the latest start time of the transmission of a frame f_i, generated at time t, of flow τ_i at its last visited output $last_i$. It is computed by the following equation:

$$W_{i,t}^{last_i} = D_{hp^X}(i,t) + D_{X_{max}}(i,t) + D_{lp^X}(i) \tag{1.3}$$

$$+ \sum_{h \in \mathcal{P}_i \backslash \{last_i\}} \left(\max_{\substack{j \in \{1,2,\dots,n\} \\ h \in \mathcal{P}_j}} (C_j)\right) \tag{1.4}$$

$$+ (|\mathcal{P}_i| - 1) \cdot sl \tag{1.5}$$

$$- \left(\sum_{h \in \mathcal{P}_i / \{first_i\}} \Delta_{i,t}^h - t\right)^+ \tag{1.6}$$

$$- C_i \cdot \left(1 + \max_{h \in \mathcal{P}_i} \left(\frac{\alpha_X^-(h)}{\alpha_X^+(h)}\right)\right) \tag{1.7}$$

Each term in the computation is briefly explained in the following paragraphs. Reader can refer to [LI 16] for details.

Term [1.3] corresponds to the sum of delay introduced by higher priority class flows, delay of flows from the same class scheduled by FIFO and non-preemptive delay introduced by lower priority flows. Precisely speaking, $D_{hp^X}(i,t)$ represents the delay of higher priority class flows. The frames of these flows can delay the studied frame f_i if they arrive no later than the start transmission time of frame f_i. Once frame f_i starts its transmission on a link, it cannot be preempted any more by a later arriving higher priority class frame. Note that even the higher priority flows are served first when there is available bandwidth; they are still constrained by the CBSA which means that their transmission cannot exceed the reserved portion of bandwidth. This point has been taken into account in the computation of term $D_{hp^X}(i, t)$ in order to lead to a higher delay upper bound for class X flows.

Term $D_{X_{max}}(i, t)$ represents the delay caused by other same-priority class X flows based on FIFO scheduling policy. Except the transmission delay, the same-priority class X flows share the same credit and are then constrained by the same CBSA traffic shaper. The waiting time imposed by CBSA traffic shaper when the credit is negative is integrated in the computation of term $D_{X_{max}}(i, t)$ by the term 1.1. Meanwhile, this waiting time can happen during the same period when higher priority class flows are transmitted, which is already considered in term $D_{hp^X}(i, t)$. Therefore, an improvement taking into account this point is also implemented in the term $D_{X_{max}}(i, t)$.

Term $D_{lp^X}(i)$ considers the non-preemptive delay introduced by one lower priority class frame at each output port along the path of flow τ_i. In order to compute the delay upper bound, the transmission time of the largest lower priority class frame is counted in the computation.

Term [1.4] represents the transmission delay of a frame sequence including frame f_i on links in the path $\mathcal{P}_i$. A frame sequence happens when several frames are transmitted one after another without interruption on a link. The transmission delay of such a frame sequence is maximized by considering the transmission time of the largest frame of the sequence.

Term [1.5] corresponds to the sum of switching latencies along the path $\mathcal{P}_i$ which includes $(|\mathcal{P}_i| - 1)$ switches. The switching latency sl at each switch is considered as an upper bounded constant.

Term [1.6] refers to the serialization term that takes into account the fact that frames transmitted on the same link are serialized, transmitted one by one, and they cannot arrive at the following switch output port at the same time. This is a physical constraint. Term $\Delta_{i,t}^h$ is the serialization term of frame f_i at the output port h. All the serialization terms at each output port, except the first one, are considered in the computation. Reader can refer to [BAU 10, BAU 12, LI 14, LI 16] for details.

Term [1.7] is subtracted since $W_{i,t}^{last_i}$ is the start time of frame f_i transmission at the output port $last_i$.

1.3.3.2.3. Case study

A case study was carried out on a frame f_1 of flow τ_1 following a path of $\mathcal{P}_1 = \{ECU_1, S_{11}, S_{21}\}$ in Figure 1.4. The transmission rate of the network is $R = 100\ Mbit/s$. Frame f_1 is generated at time t at its source node $first_1 = ECU_1$. The flow temporal characteristics are given in Table 1.1, which leads to three flow index sets of class B flow τ_1: $hp^B = \{4\}$, $classB = \{1, 2, 3\}$ and $lp^B = \{5\}$. All flows in the example are periodic flows. For simplicity reasons, the switching latency is considered null, i.e. $sl = 0$.

τ_i	$T_i\ (\mu s)$	$C_i\ (\mu s)$	Pr_i	AVB class
τ_1	2000	40	2	class B
τ_2	2000	40	2	class B
τ_3	2000	40	2	class B
τ_4	160	40	3	class A
τ_5	2000	40	1	class C
τ_6	2000	120	1	class C

Table 1.1. *Flow temporal parameters of an AVB switched Ethernet network example*

The SRP is disabled in the example and fixed set values of *IdleSlope* and *SendSlope* for each AVB SR class: the AVB SR class A with $\alpha_A^+(h) = \alpha_A^-(h) = 0.5 \times R$ as well as the AVB SR class B with $\alpha_B^+(h) = 0.25 \times R$ and $\alpha_B^-(h) = 0.75 \times R$ for each output port h. For simplicity reasons, we use α_X^+ and α_X^- in the following paragraphs to represent the traffic shaping characteristics for AVB SR classes X (A or B). The total bandwidth reserved by the two AVB SR classes is 75%, respecting the recommended requirement for AVB SR classes.

For illustration purposes, the scenario of $t = 0$ is considered. The workload introduced by all class B flows $classB = \{1, 2, 3\}$ is computed by:

$$D_{B_{max}}(1, t = 0) = 480\ \mu s$$

A class A flow τ_4, a higher priority class flow, joins the path $\mathcal{P}_1$ at the output port S_{21}. As illustrated in Figure 1.6, three frames f_4, f_4' and f_4'' of flow τ_4 could delay frame f_1. The delay introduced by flow τ_4 is computed by:

$$D_{hp^B}(1, t = 0) = 120\ \mu s$$

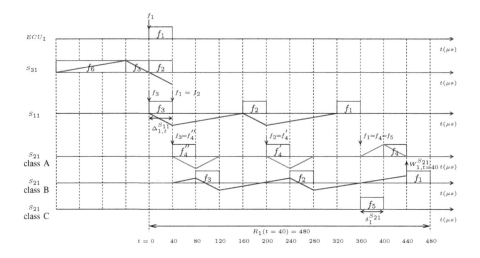

Figure 1.6. *The scenario of frame f_1 with AVB switching*

Since the transmission time of frames f_4' and f_4'' happens when the credits of class B is replenished, the maximum workload introduced by all class B flows is refined to the following value:

$$D_{B_{max}}(1, t = 0) = 400 \, \mu s$$

Based on the computation of serialization term, the following results are given:

$$\Delta_{1,t=0}^{S_{11}} = \Delta_{1,t=0}^{S_{21}} = 0$$

Then, the serialization term is computed by:

$$\left(\sum_{h \in \{S_{11}, S_{21}\}} \Delta_{1,t=0}^{h} - t \right)^+ = 0$$

There is only one lower priority flow τ_5 delaying flow τ_1 at the output port S_{21}, and then the non-preemptive delay is $D_{lp^B}(1) = C_5 = 40 \, \mu s$. As flows τ_i ($i \in \{1, 2, \ldots, 5\}$) have the same frame transmission time $C_i = 40 \, \mu s$, the transmission delay along the path $\mathcal{P}_1$ is then $80 \, \mu s$. Hence, the latest start time of the transmission of frame f_1 at the output port S_{21} is bounded by:

$$W^{S_{21}}_{1,t=0} = D_{A_{max}}(1, t = 0) + D_{B_{max}}(1, t = 0) + 40$$

$$-(\sum_{h \in \{S_{11}, S_{21}\}} \Delta^h_{1,t=0} - t)^+ + 80 - 40 \cdot (1 + \frac{\alpha^-_B}{\alpha^+_B})$$

$$= 480 \, \mu s \qquad\qquad\qquad [1.8]$$

which lead to the delay of flow τ_1 as follows:

$$R_1 = W^{S_{21}}_{1,t=0} + 40 - t = 520 \, \mu s$$

In order to guarantee the maximum delay, we check each possible value of t ($0 \le t \le \mathcal{B}_1$) with $\mathcal{B}_1 = 640 \, \mu s$ by equation [1.2]. In the example, the maximum delay of flow τ_1 is $R_1 = 520 \, \mu s$, which is obtained when $t = 0$.

One scenario of frame f_1 is given in Figure 1.6. The blue lines represent the class B credit, while the red lines represent the class A credit. The downward arrows represent the arrivals of frames, while $f_i = f_j$ means that frame f_i arrives at the same time as frame f_j. It indicates that the worst-case delay of frame f_1 is $480 \, \mu s$, which means that the computed result of $520 \, \mu s$ is pessimistic. Actually, there does have a serialization term of $\Delta^{S_{11}}_{1,t=0} = 40 \, \mu s$ as illustrated in Figure 1.6. Hence, there is an introduced pessimism of $40 \, \mu s$. In a general case, where it is difficult to exactly verify the existence of the interval of size $I^h_{X_{init}}$, we suppose that it always exists in order to calculate the maximum delay. Therefore, this introduces some pessimism in some scenarios.

1.4. A representative automotive configuration

A representative automotive configuration based on AVB switching and its real-time performances are shown in the following. First, an automotive application example for a representative network configuration is introduced. Then, the end-to-end delays are studied based on a comparison between the results obtained by simulations and the ones obtained by AVB Trajectory approach.

The network example is illustrated in Figure 1.7. It is composed of eight ECUs interconnected by two switches via full-duplex links of 100 $Mbit/s$.

The switching latency is 8 μs for each switch. This network example is built based on the results published in [LIM 11, STE 12, ALD 12].

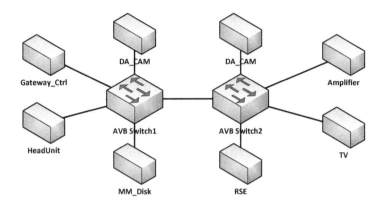

Figure 1.7. *An automotive AVB switched Ethernet network example*

The automotive application studied here consists of advanced driver-assistance system (ADAS), a multimedia application and an infotainment system. The streams from the ADAS as well as from the control system are class A flows, while the streams from the multimedia and infotainment system are class B flows. The Gateway_Ctrl ECU is a device that connects the studied network to the traditional control network and transmits cross-domain control traffic to the control processing unit (Ctrl_PU) and the rear seat entertainment (RSE) system. The driving assistance camera (DA_CAM) ECU is a device that processes data of cameras installed in different parts inside a vehicle (e.g. front camera, back camera, side cameras, etc.). DA_CAM ECU then aggregates camera streams and sends it to the head unit (HeadUnit) in order to display the needed information, e.g. a warning or back view when reversing the vehicle. Multimedia disk (MM_Disk) player sends audio/video streams to the amplifier (audio) and the RSE (video). Then, the live TV data is streamed from the TV ECU to the HeadUnit. The characteristics of each flow transmitted in the example network are given in Table 1.2. All the flows are periodic. For the purpose of performance comparison, the end-to-end delay of each flow is computed by both the OMNet++ [OMN 14] simulation tool and the AVB trajectory approach extension. We consider the simulation model developed in [STE 11] with support of the INET-Framework [INE 10].

τ_i	Source ECU	Destination ECU	T_i (μs)	Payload (byte)	C_i (μs)	Pr_i	AVB class
τ_1	Gateway_Ctrl	Ctrl_PU	5000	46	5.12	3	class A
τ_2	Gateway_Ctrl	RSE	5000	46	5.12	3	class A
τ_3	DA_CAM	HeadUnit_Ctrl	125	390	32.64	3	class A
τ_4	MM_Disk	RSE	280	1428	115.68	2	class B
τ_5	MM_Disk	Amplifier	1400	1428	115.68	2	class B
τ_6	TV	HeadUnit_Ctrl	560	1428	115.68	2	class B

Table 1.2. *Flow temporal parameters of the AVB automotive example*

The simulations were executed 200 times on duration equal to five times the least common multiple (LCM) of all flow periods and each time all the traffic started at a random time. The average and maximum observed delays obtained by simulation for each flow are listed in columns *Sim. mean* and *Sim. max*, respectively, in Table 1.3. The maximum observed delay can be obtained in a real network. However, it may not be the worst-case delay since simulations cannot guarantee that the frames experienced the worst-case scenario.

τ_i	Sim. mean (μs)	Sim. max (μs)	Traj. approach (μs)	Pess. upper bound (%)
τ_1	327.6	1819.6	2654.6	31.4%
τ_2	130.0	1941.8	2765.2	29.8%
τ_3	88.9	192.0	194.7	1.4%
τ_4	384.7	568.3	610.1	6.7%
τ_5	447.3	532.2	610.1	12.8%
τ_6	369.8	399.3	403.4	1.0%

Table 1.3. *The optimism evaluation of the AVB trajectory extension*

The results show a great gap between the average delay and the maximum delay of class A flows. This gap is reduced for class B flows. The comparison of the average delay and the maximum delay indicates the potential over-dimensional worst-case scenario, especially for class A flows.

The same network example has been considered with the extended trajectory approach for the AVB switching. The worst-case end-to-end delay bound of each flow obtained by the trajectory approach is also listed in Table 1.3. This obtained value is an upper bound on end-to-end delay, which cannot be exceeded. Comparing the maximum delays obtained by simulations to those obtained by the trajectory approach, we can bound the maximum pessimism introduced by the AVB trajectory approach. The *Pess. upper bounds* are given in Table 1.3. It can be seen that the trajectory approach can introduce at most 31.4% (τ_1) of pessimism compared with the results obtained by simulations for the studied network example.

Exact worst-case end-to-end delays in a large-scale network cannot be obtained with a formal method approach like model-checking due to state-space explosion problem of exhaustive computation. Hence, the trajectory approach provides safe upper bounds that can be used for performance evaluation of an AVB switched Ethernet network.

1.5. Conclusion

Switched Ethernet AVB standards and the current Ethernet TSN evolution are a set of standards that make Ethernet solutions valid for audio, video and real-time communications. It is currently being put to use in the automotive industry and can be seen as a mature solution. In this chapter, we have presented physical and medium access layers for Ethernet and showed with representative examples that it is compatible with most of worst-case end-to-end delay constraints associated with automotive flows, including audio and video communications.

1.6. Bibliography

[ALD 12] ALDERISI G., CALTABIANO A., VASTA G. *et al.*, "Simulative assessments of IEEE 802.1 Ethernet AVB and time-triggered Ethernet for advanced driver assistance systems and in-car infotainment", *2012 IEEE Vehicular Networking Conference (VNC)*, pp. 187–194, 2012.

[BAU 10] BAUER H., SCHARBARG J.-L., FRABOUL C., "Improving the worst-case delay analysis of an AFDX network using an optimized trajectory approach", *IEEE*, vol. 6, no. 4, pp. 521–533, 2010.

[BAU 12] BAUER H., SCHARBARG J.-L., FRABOUL C., "Applying trajectory approach with static priority queuing for improving the use of available AFDX resources", *Real-time Systems*, vol. 48, no. 1, pp. 101–133, January 2012.

[BOR 14] BORDOLOI U., AMINIFAR A., ELES P. *et al.*, "Schedulability analysis of Ethernet AVB switches", *2014 IEEE 20th International Conference on Embedded and Real-Time Computing Systems and Applications (RTCSA)*, pp. 1–10, 2014.

[DIE 12a] DIEMER J., ROX J., ERNST R., "Modelinig of ethernet AVB networks for worst-case timing analysis", *IFAC Proceedings Volumes*, vol. 45, no. 2, pp. 848–853, 2012.

[DIE 12b] DIEMER J., THIELE D., ERNST R., "Formal worst-case timing analysis of ethernet topologies with strict-priority and AVB switching", *7th IEEE International Symposium on Industrial Embedded Systems (SIES)*, pp. 1–10, 2012.

[IEE 00] IEEE, IEEE Standard for Carrier Sense Multiple Access with Collision Detection (CSMA/CD) Access Method and Physical Layer Specifications, available at: https://standards.ieee.org/findstds/standard/802.3-2000.html, 2000.

[IEE 09] IEEE, IEEE Standard for Local and Metropolitan Area Networks – Virtual Bridged Local Area Networks – Amendment: Forwarding and Queuing Enhancements for Time-Sensitive Streams, available at: http://odysseus.ieee.org/query.html, 2009.

[IEE 10] IEEE, IEEE Standard for Local and Metropolitan Area Networks – Virtual Bridged Local Area Networks – Amendment: 9: Stream Reservation Protocol (SRP), available at: http://odysseus.ieee.org/query.html, 2010.

[IEE 11a] IEEE, IEEE Standard for Local and Metropolitan Area Networks – Audio/Video Bridging (AVB) Systems, available at: http://odysseus.ieee.org/query.html, 2011.

[IEE 11b] IEEE, IEEE Standard for Local and Metropolitan Area Networks – Timing and Synchronization for Time-Sensitive Applications in Bridged Local Area Networks Amendment: Enhancements and performance improvements, available at: http://odysseus.ieee.org/query.html, 2011.

[IEE 11c] IEEE, IEEE Standard for Virtual LANs, available at: http://odysseus.ieee.org/query.html, 2011.

[IEE 12] IEEE, 802.3-2012 – IEEE Standard for ethernet, Standard 1516–2000, pp. 1–3747, 2012.

[IMT 09] IMTIAZ J., JASPERNEITE J., HAN L., "A performance study of ethernet audio video bridging (AVB) for industrial real-time communication", *Proceedings of the 14th IEEE International Conference on Emerging Technologies & Factory Automation*, IEEE Press, pp. 1133–1140, 2009.

[INE 10] INET, INET Framework, available at: https://inet.omnetpp.org/News.html, 2010.

[LI 14] LI X., CROS O., GEORGE L., "The trajectory approach for AFDX FIFO networks revisited and corrected", *IEEE 20th International Conference on Embedded and Real-Time Computing Systems and Applications (RTCSA)*, IEEE, pp. 1–10, 2014.

[LI 16] LI X., GEORGE L., "Deterministic delay analysis of AVB switched ethernet networks using an extended trajectory approach", in *Real-Time Systems*, Springer, 2016.

[LIM 11] LIM H.-K., VÖLKER L., HERRSCHER D., "Challenges in a future IP/ethernet-based in-car network for real-time applications", *Proceedings of the 48th Design Automation Conference*, ACM, pp. 7–12, 2011.

[LIM 12] LIM H.-T., HERRSCHER D., CHAARI F., "Performance comparison of IEEE 802.1Q and IEEE 802.1 AVB in an ethernet-based in-vehicle network", *8th International Conference on Computing Technology and Information Management (ICCM)*, vol. 1, pp. 1–6, April 2012.

[MAR 05] MARTIN S., MINET P., GEORGE L., "End-to-end response time with fixed priority scheduling: trajectory approach versus holistic approach", *International Journal of Communication Systems*, vol. 18, no. 1, pp. 37–56, Wiley Online Library, 2005.

[MAR 06a] MARTIN S., MINET P., "Schedulability analysis of flows scheduled with FIFO: application to the expedited forwarding class", *Proceedings of International Parallel and Distributed Processing Symposium (IPDPS)*, Rhodes Island, Greece, IEEE, p. 8, April 2006.

[MAR 06b] MARTIN S., MINET P., "Worst case end-to-end response times of flows scheduled with FP/FIFO", *Proceedings of International Conference on Networking*, Mauritius, pp. 54–60, April 2006.

[MAR 06c] MARTIN S., MINET P., GEORGE L., "The trajectory approach for the end-to-end response times with non-preemptive FP/EDF*", *Software Engineering Research and Applications*, Springer, pp. 229–247, 2006.

[OMN 14] OMNeT, OMNeT++ 4.6, available at: https://omnetpp.org/9-articles/software/3724-omnet-4-6-released, 2014.

[REI 13] REIMANN F., GRAF S., STREIT F. *et al.*, "Timing analysis of ethernet AVB-based automotive E/E architectures", *IEEE 18th Conference on Emerging Technologies & Factory Automation (ETFA)*, IEEE, pp. 1–8, 2013.

[STE 08] STEINER W., "TTEthernet specification", *TTTech Computertechnik AG*, November, 2008.

[STE 11] STEINBACH T., KENFACK H.D., KORF F. *et al.*, "An extension of the OMNeT++ INET framework for simulating real-time ethernet with high accuracy", *Proceedings of the 4th International ICST Conference on Simulation Tools and Techniques*, pp. 375–382, 2011.

[STE 12] STEINBACH T., LIM H.-K., KORF F. *et al.*, "Tomorrow's in-car interconnect? A competitive evaluation of IEEE 802.1 AVB and time-triggered ethernet (AS6802)", *Vehicular Technology Conference (VTC Fall), 2012 IEEE*, IEEE, pp. 1–5, 2012.

[THI 16] THIELE D., ERNST R., "Formal worst-case timing analysis of ethernet TSN's burst-limiting shaper", *2016 Design, Automation Test in Europe Conference Exhibition (DATE)*, pp. 187–192, March 2016.

[TIN 94] TINDELL K., CLARK J., "Holistic schedulability analysis for distributed hard real-time systems", *Microprocessing and Microprogramming*, vol. 40, no. 2, pp. 117–134, Elsevier, 1994.

[TUO 15] TUOHY S., GLAVIN M., HUGHES C. *et al.*, "Intra-vehicle networks: a review", *IEEE Transactions on Intelligent Transportation Systems*, vol. 16, no. 2, pp. 534–545, IEEE, 2015.

Representation of Networks of Wireless Sensors with a Grayscale Image: Application to Routing

2.1. Introduction

Wireless sensor networks (WSN) have some specific constraints. One major constraint is due to the fact that sensors are deployed in a hostile area and that they need to be powered by batteries. Therefore, the lifetime of the sensor is dramatically related to the amount of initial power loaded in the battery and to the way it is consumed. The control of the energy consumption of sensors is an active research field [YI 11].

The energy consumption of a sensor is used for data acquisition, processing, formatting and transmission. Data transmission is the primary energy consumption, hence we will focus our study on the protocols for the distribution of communication energy. These protocols can act on several layers of communications: the physical layer, MAC or access layer, network or routing layer and application layer.

On the access layer, energy is wasted by packet retransmission after collision, control packet transmission without useful data or "processing of packets to reach other sensors". Some protocols are proposed to act on this waste of energy. The S-MAC protocol [AMM 10], for example, uses time

Chapter written by Marc GILG.

synchronization between sensors to idle them cyclically. Another protocol using synchronization is presented in [ABD 12], and some protocols also use traffic information to save energy [ALA 11]. The PC-MAC protocol [HU 11] avoids collisions to preserve energy.

In the network layer, routing protocols will minimize energy consumption by the way they deliver packets. These routing protocols are also affected by the hardware limitation of sensors, especially the lack of memory and processing capabilities. On the other hand, they will try to take advantage of the high density of wireless sensor networks. Creating new wireless sensor network routing protocols is a popular challenge. They can be classified by the way they operate [GOY 12]. Routing protocols can be geocentric, data-centric or use network topology and link states. In data-centric protocols, the GKAR [WAN 12b] is an example of a K-anycast protocol. EASPRP [DOO 12] looks for the shortest path with energy constraints and the EERT protocol [MAZ 12] takes care of quality of service and energy. In real-time networks, REFER protocol [LI 12] uses the Kautz graphs. Some routing protocols can be used together to cover an area well [PAT 12].

Some routing protocols [FAL 12] use the moving capacity of mobile sensors. The GAROUTE protocol [SAR 11] uses the genetic algorithm to create clusters of sensors to reduce communications. Cluster heads can be used to process data locally and to aggregate them [ZHA 12, TOY 12, SHI 12]. LEACH protocol [SHA 12, CHE 12, HAN 11] was the first protocol to use clusters to balance network traffic. This protocol was modified in [WAN 12a] to improve delivery time and to minimize interferences. EEDR protocol [HUA 12] is focused on packet forwarding to minimize energy consumption.

Other methods were proposed to reduce energy consumption and to increase lifetime of wireless sensor networks. In [HUA 11], sensor coverage is used to minimize energy consumption. This is also done in [ZHU 12] which used the SCC (Sponsored Coverage Calculation) simulation method. Other protocols minimize data transmissions like the SEPSen protocol [KAS 12] which uses data processing along the way. Traffic and resource management is used in [MAD 12]. Most of these protocols use wireless network simulators [LAH 12].

In this chapter, a new method using a gray level image analogy of wireless sensor networks is introduced. This method can be found in

[GIL 09, YOU 10b, YOU 10a]. A grayscale image is constructed, where each pixel represents a sensor and its gray level represents the battery capacity of this sensor. On this representation of the network, different image processing algorithms are used to produce routing algorithms for the network.

2.2. The image analogy

2.2.1. *Gray level images*

A digital image is made of a set of points, called pixels, located on integer coordinates. Each pixel has a more or less intense brightness. The juxtaposition of these pixels results in a grayscale picture, like black and white televisions, see Figure 2.1.

Figure 2.1. *Gray level image*

On such a picture, a human can recognize shapes, objects, people, etc. For a computer, this is not obvious and lots of algorithms have been developed for this task, such as border detection algorithms, character recognition, and so on. It is important to note that a pixel is usually surrounded by eight other pixels.

The brightness or gray level is represented by a numerical value taken at a certain interval. The greater the range of shades in the image, the larger this interval is. In computers, the processing is done on digital values 0 or 1. A brightness encoded in 8 bits will take 2^8 possible gray levels for 0 to 255. By convention, black will be 0 and white 255.

2.2.2. *Construction of a gray level image for a network*

A gray level image can be used to represent the geographical distribution of a value. In a wireless sensor network, communication is made between sensors. On each sensor, this communication will be characterized by different parameters: the number of packets to be sent or received, the signal power, and so on. The remaining battery capacity is also a value given by each sensor. These values are continuous, but can be digitized and represented by an integer. If this integer is between 0 and 255, this means that it is an 8 bit encoding.

Now suppose that each sensor is located on a grid with coordinates (x, y) and has an 8 bit value for some parameters, say battery capacity for example. This can be seen as an 8 bit gray level image, where each pixel is a sensor and the brightness is the battery capacity, as shown in Figure 2.2.

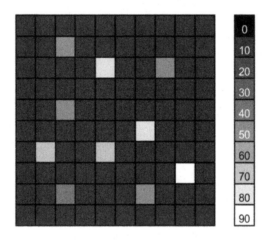

Figure 2.2. *10 sensors network on a gray level image*

In Figure 2.2, sensors in the same communication range are not necessarily neighbors. This can be troublesome if we wish to apply image processing algorithms, such as a convolution filter, for example. To avoid this, eight virtual neighbors will be created, where the gray level given to neighbors will be the average gray level of sensors in communication range.

2.2.3. *Virtual neighbors*

In an image, a pixel usually has eight neighbors. In a network, the notion of neighbor is defined to be in the communication range. In the gray level image analogy, an image processing algorithm, like the convolution filter, is applied to a sensor and its neighbors. A virtual neighbor represents a pixel with a gray level defined by the mean value of the sensors in communication range.

The virtual neighbors around a sensor N are defined by these rules:

– consider only the sensors in communication range of N;

– gather these sensors in eight sectors around the sensor N;

– a gray level is defined for each sector by computing the mean value of the sensors in the sector;

– the eight sectors with the gray level are considered as the eight virtual neighbors of sensor N.

Figure 2.3 shows the eight sectors around the sensor N_0.

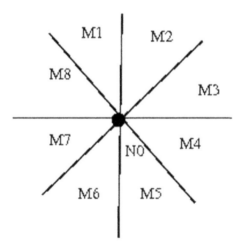

Figure 2.3. *The eight sectors around the sensor N_0*

A local image representation can be defined as a matrix Mv, where the entries $M_1, \ldots, M_8$ represent the gray level of the eight virtual neighbors:

$$Mv = \begin{pmatrix} M_8 & M_1 & M_2 \\ M_7 & N_0 & M_3 \\ M_6 & M_5 & M_4 \end{pmatrix} \qquad [2.1]$$

The matrix [2.1] can be used for image processing algorithms like matrix convolution processing.

2.3. Image processing algorithm

2.3.1. Convolution filters

Numerous image processing techniques exist. One technique, the convolution filter, consists of replacing the brightness of a pixel with a brightness value computed with the eight neighbors brightness value. This filter uses several types of kernel: the Gaussian kernel [BAS 02] or Sobel kernel [JIN 09, CHU 09, JIA 09, BAB 03], for example.

An image can be seen as a matrix I, where $I(x, y)$ is the brightness of the pixel located at coordinates (x, y). A convolution product is computed between the matrix I and a kernel matrix K which represents the type of filter. K can be of size 3×3 or 5×5. The result of this product will be the new brightness of the pixel (x, y).

The convolution product $*$ for a kernel of size 3×3 is defined by:

$$I * K = \begin{pmatrix} I(1,1) & I(1,2) & \ldots & I(1,n) \\ \vdots & & I(x,y) & \vdots \\ I(m,1) & I(m,2) & \ldots & I(m,n) \end{pmatrix}$$
$$* \begin{pmatrix} K(1,1) & K(1,2) & K(1,3) \\ K(2,1) & K(2,2) & K(2,3) \\ K(3,1) & K(3,2) & K(3,3) \end{pmatrix} \qquad [2.2]$$

where

$$I * K_{x,y} = \sum_{i=-1}^{1} \sum_{j=-1}^{1} I(x+i, y+j) * K(2+i, 2+j) \qquad [2.3]$$

The choice of kernel K induces the type of filter used.

2.3.2. *Mean filter*

The mean filter is a smoothing filter that replaces the brightness by the mean of the brightness of the pixel itself and its neighbors. The mean filter uses the kernel:

$$K = \begin{pmatrix} \frac{1}{9} & \frac{1}{9} & \frac{1}{9} \\ \frac{1}{9} & \frac{1}{9} & \frac{1}{9} \\ \frac{1}{9} & \frac{1}{9} & \frac{1}{9} \end{pmatrix}$$

Figure 2.4 is an example of a mean filter on a photograph.

Figure 2.4. *Mean filter*

2.3.3. *Gradient filter*

A gradient vector represents the value's variation in a certain direction. Here, the gradient filter gives the brightness variation in direction X or Y. The kernel K depends on the direction and is defined as:

$$\frac{\partial I}{\partial x} = \begin{pmatrix} 0 & 1 & -1 \\ 0 & 1 & -1 \\ 0 & 1 & -1 \end{pmatrix} \qquad [2.4]$$

$$\frac{\partial I}{\partial(-x)} = \begin{pmatrix} -1 & 1 & 0 \\ -1 & 1 & 0 \\ -1 & 1 & 0 \end{pmatrix}$$ [2.5]

$$\frac{\partial I}{\partial(-y)} = \begin{pmatrix} 0 & 0 & 0 \\ 1 & 1 & 1 \\ -1 & -1 & -1 \end{pmatrix}$$ [2.6]

$$\frac{\partial I}{\partial y} = \begin{pmatrix} -1 & -1 & -1 \\ 1 & 1 & 1 \\ 0 & 0 & 0 \end{pmatrix}$$ [2.7]

Figures 2.5 and 2.6 show a filtered image in X and Y directions.

Figure 2.5. *X gradient filter*

2.3.4. *Gaussian filter*

The Gaussian filter uses the Gaussian function in the kernel of the filter:

$$G(x, y) = \frac{1}{2\pi\sigma^2} \exp^{\frac{x^2+y^2}{2\sigma^2}}$$

This filter is a weighted filter which gives more importance to the central pixels. The parameter σ controls the weight given to the center.

Figure 2.6. *Y gradient filter*

Figure 2.7 is an example of the Gaussian filter.

Figure 2.7. *Gaussian filter with $\sigma = 0.04$*

2.3.5. *Sobel filter*

The Sobel filter is a border detection filter. It uses two gradient kernels G_x and G_y. Each pixel gets a new gray value given by $\sqrt{G_x^2 + G_y^2}$. The kernels G_x and G_y are defined by:

$$G_x = \begin{pmatrix} -1 & -2 & -1 \\ 0 & 0 & 0 \\ 1 & 2 & 1 \end{pmatrix} \qquad\qquad [2.8]$$

$$G_y = \begin{pmatrix} -1 & 0 & 1 \\ -2 & 0 & 2 \\ -1 & 0 & 1 \end{pmatrix} \qquad\qquad [2.9]$$

Figures 2.8 and 2.9 are examples of the actions of G_x and G_y.

Figure 2.8. G_x *gradient filter*

The Sobel filter gives the border of Figure 2.10. The gray level is inverted to get better readability.

Some other convolution filters can be defined, but they are not used in this chapter.

Figure 2.9. G_Y *gradient filter*

Figure 2.10. *Sobel filter example (inverted gray level)*

2.3.6. *Deformable models for border detection*

Border detection is a pertinent way to separate regions covered by sensors of high and low energy level. The Sobel filter, in section 2.3.5, is one way to identify these borders, but there exists another method.

In [MCI 96], deformable curves are used by D. Terzopoulos to find borders in a medical image.

A curve on a plan surface of coordinates $(x, y) \in \mathbb{R}^2$ can be defined by a parametric function $v(s) = (x(s), y(s))^T$, where $s \in [0, 1]$ is its curvilinear abscissa. The shape of this curve is given by the forces applied on it:

$$\mathcal{E}(v) = S(V) + \mathcal{P}(v) \qquad [2.10]$$

The function $\mathcal{E}$, which represents the energy of the curve, is the sum of the deformation energy S used to give the shape of the curve and $\mathcal{P}$ the potential energy induced by the image. The curve will minimize the energy $\mathcal{E}$.

2.3.6.1. Energy for deformation

The energy for deformation of a curve S is the energy needed to modify the shape of the curve. This energy depends on the tension w_1 and the rigidity w_2 of the curve and it is given by:

$$S(v) = \int_0^1 w_1(s) \left| \frac{\partial v}{\partial s} \right|^2 + w_2(s) \left| \frac{\partial^2 v}{\partial s^2} \right|^2 ds \qquad [2.11]$$

Increasing w_1 increases tension, which results in sharp loops, increasing w_2 gives more rigidity, and the loops will be smoother.

2.3.7. Potential energy

The potential energy $\mathcal{P}$ is linked to the brightness of the image by a scalar function P. The potential energy is defined by:

$$\mathcal{P}(v) = \int_0^1 P(v(s))ds \qquad [2.12]$$

The scalar function P can be defined, for example, using the Gaussian filter:

$$P(x, y) = -c|\nabla[G_\sigma \times I(x, y)] \qquad [2.13]$$

where c is a positive number, G_σ the kernel of the Gaussian filter of parameter σ and ∇ the gradient operator.

2.3.8. *Minimal energy curve*

The curve v that minimizes the energy $\mathcal{E}(v)$ is a solution of equation:

$$-\frac{\partial}{\partial s}\left(w_1\frac{\partial v}{\partial s}\right) + \frac{\partial^2}{\partial s^2}\left(w_2\frac{\partial^2 v}{\partial s^2}\right) + \nabla P(v) = 0 \qquad [2.14]$$

Equation [2.14] shows that the minimal energy $\mathcal{E}(v)$ is given by the equilibrium of the energy of deformation and the potential energy. Equation [2.14] is not easy to solve, a solution can be given by a discrete numerical approximation.

2.3.9. *Discretion*

Equation [2.10] can be discretized as:

$$E(\mathbf{u}) = \frac{1}{2}\mathbf{u}^T\mathbf{Ku} + \mathbf{P(u)} \qquad [2.15]$$

where $\mathbf{u}$ represents a sequence of vectors which models the curve v, K a rigidity matrix of the curve and P a discrete representation of the potential energy. Equation 2.15 will be used to construct routing algorithms.

2.4. Routing algorithms

In this section, three routing algorithms based on grayscale are presented. Two of them use convolution filters: one is based on the Sobel filter and one on the meaning filter. The third one is based on route deformation and the energy curve $E(v)$ given by equation 2.15.

2.4.1. *Routing algorithm based on the Sobel filter*

The routing algorithm that is proposed here will modify the AODV routing algorithm to forward the packets to high energy level regions. These regions are identified by gradients computed using the Sobel filter. This algorithm is taken from the paper [GIL 09]. This protocol has two steps: computation of the energy gradient and routing using modified AODV protocol.

2.4.1.1. *Gradient computation*

In this first step, a sensor will use the kernels G_x and G_y, defined in equations [2.8] and [2.9], to compute its energy gradient. To do this, the matrix Mv [2.1] defined by the energy sectors is used.

The sensor runs the following computation algorithm, where C represents the energy capacity left in its battery:

1) If the sensor has little energy, say $C < C_0$, then it does not participate in the routing process.

2) The sensor collects its neighbors battery capacity C_i and their position.

3) The sensor computes the matrix Mv from the information it collects.

4) The sensor computes the gradient (g_x, g_y) with $g_x = -G_y * Mv$ and $g_y = -G_x * Mv$.

After this computation, the sensor has a gradient vector $G = (g_x, g_y)$ pointing to the direction of the high energy region. This vector will be used as the modified AODV routing algorithm.

2.4.1.2. *Routing protocol*

The AODV routing protocol (*Ad hoc* On-demand Distance Vector) is made for ad-hoc networks. It uses message broadcasting to discover the route from source to destination.

The AODV protocol will perform these steps:

1) The source broadcasts RREQ messages.

2) If a node receives the RREQ message and it is not the destination, it forwards the message to its neighbors. It keeps the source node and the last transmitter of the message in its routing table.

3) If the destination node gets the RREQ message, it answers with the RREP message.

4) If a node gets an RREP message, it forwards the message to the source.

5) If the source gets an RREP message, it chooses the shortest path to the destination from its routing table.

The last step of this protocol will be modified. The path closer to the energy gradient will be kept. To be close means that the angle between the gradient vector and the vector V pointing from the current node to the next node in the path is minimum. The cosine of this angle is given by the scalar product:

$$\cos(V, G) = \frac{V \times G}{\|V\| \times \|G\|} \qquad [2.16]$$

For each path, the RREP message will keep the minimal value of this cosine. As the variation of the angle is opposite to the variation of the cosine, the RREP message will keep the greatest deviation between the path and the energy gradient. Finally, the route chosen will be the route with maximal cosine value kept in RREP. This means that the route crossing the highest energy region will be chosen.

2.4.1.3. Application

The routing protocol with Sobel kernels is used in the following network:

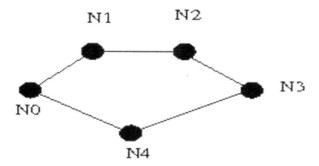

Figure 2.11. *Four-sensor network*

The source node is N_0 and the destination node is N_3. Two paths are possible: $P_1 = (N_0, N_1, N_2, N_3)$ and $P_2 = (N_0, N_4, N_3)$. Two scenarios are compared. In the first one, all the nodes have the same initial energy capacity, say 10 for example. In the second scenario, the node N_4 has half the energy capacity, say 5.

The gradient vector can be computed for each node in the two scenarios:

– Scenario 1:

	G_0	G_1	G_2	G_3	G_4
x	20	10	−10	−20	0
y	0	10	10	0	−20

– Scenario 2:

	G_0	G_1	G_2	G_3	G_4
x	15	10	−10	−15	0
y	−5	10	10	−5	−20

For each path, the vectors pointing in the same direction can be computed:

P_1	N_0	N_1	N_2
x	−1	−1	−1
y	1	0	−1

P_2	N_0	N_4
x	−1	−1
y	−1	1

From this information, the cosine can be computed:

– Scenario 1:

P_1	N_0	N_1	N_2
cos	−0.7	−0.7	0

P_2	N_0	N_4
cos	−0.7	−0.7

The maximal absolute value of the cosine is 0.7. The AODV rule is applied and the shortest path P_2 is selected.

– Scenario 2:

P_1	N_0	N_1	N_2
cos	−0.8	−0.7	0

P_2	N_0	N_4
cos	−0.4	−0.7

P_1 has an absolute value of 0.7 for its cosine and P_2 a value of 0.4. The maximal value is 0.7 and the path P_1 is selected.

In scenario 2, the sensor N_4 has less energy. The path selected is P_1 which includes N_1 and N_2, even if this path is longer than P_2. The protocol has selected the path which includes high energy capacity sensors. This choice is different from the AODV standard protocol.

2.4.2. *Routing protocol with the mean filter*

In this section, we propose a protocol using the mean filter similarly to [YOU 10b]. The mean filter is computed on a particular sensor CH, called cluster head, which is designed by the administrator. The cluster head collects the energy left in batteries from the sensor of the network via the UDP protocol. Only sensors with enough energy, above an Energy Threshold, ET, will participate in the routing process.

2.4.2.1. *Parameters*

– X_i, Y_i are the coordinates of sensor N_i;

– BC_i is the battery capacity left in the sensor N_i;

– ET is the Energy Threshold;

– $SEID$ is Sending Energy Information Delay between two control packets;

– T is the last delay when a control packet was sent (current time – transmission time);

– R is the communication range of a sensor;

– M_i is the energy matrix of sensor N_i;

– K is the mean filter kernel matrix;

– ERP_i: Energy Routing Parameter is the convolution product $K * M_i$.

The kernel K of the mean filter is given by the matrix:

$$K = \begin{pmatrix} 1/12 & 1/12 & 1/12 \\ 1/12 & 4/12 & 1/12 \\ 1/12 & 1/12 & 1/12 \end{pmatrix} \qquad [2.17]$$

2.4.2.2. *Routing protocol*

Each sensor is running the following algorithm:

1) Energy information broadcasting: If BC_i has changed and if $(T > S EID)$, then BC_i and X_i, Y_i are sent to CH by a UDP control packet.

2) Computation performed by CH:

   a) CH gets the information of all the sensors N_i.

   b) CH updates its table with the IP address of N_i and the values BC_i, X_i, Y_i.

   c) CH computes $ERP_i = K_i * M_i$.

   d) CH sends the value ERP_i to N_i.

3) On the sensors N_i:

   a) N_i gets the value ERP_i from CH.

   b) If $(ERP_i < ET)$, then N_i drops packets that are not control packets.

2.4.2.3. *Simulations*

The simulations are done with the OMNeT++ simulator. The network is made of four sensors located as in Figure 2.12. N_1 is the source sensor and N_0 the destination sensor.

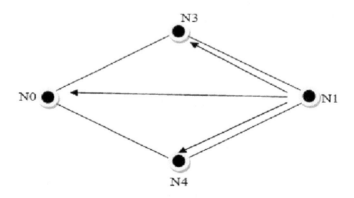

Figure 2.12. *Four-sensor network topology*

The communication protocol is 802.11 with a consumption of 250mA in transmission and 190mA in reception. When the sensor is idle, its consumption is 8mA like a Nano WiReach sensor. The traffic used is a UDP stream from N_1 to N_0 with a burst of 0.01s and a packet size of 512 bytes. The routing protocol is AODV. The initial battery capacity for the sensors N_0, N_1, N_4, N_4 will be 400, 200, 20 and 100, respectively. The results given are the mean of 10 simulations and each simulation is done for a delay of 1500s.

The following tables show the number of packets sent with the values 0.1, 0.5, 1.0 and 1.5 of $S EID$. The Energy Threshold ET will be 0, 25, 50, 75 and 100.

ET	N_0	N_1	N_3	N_4
0	14300	52500	34100	32800
25	14300	52500	9740	32200
50	14300	52500	10900	8380
75	1620	572	1200	1170
100	1600	571	1170	1180

Table 2.1. *Packets sent with $S EID = 0.1$*

ET	N_0	N_1	N_3	N_4
0	2950	48800	24700	26900
25	2950	48800	1990	28500
50	2960	48800	2100	2020
75	1070	392	691	905
100	1030	377	664	871

Table 2.2. *Packets sent with $S EID = 0.5$*

ET	N_0	N_1	N_3	N_4
0	1510	48300	25600	23700
25	1510	48300	1080	24500
50	1510	48300	1060	1110
75	237	93	203	158
100	237	93	203	158

Table 2.3. *Packets sent with $S EID = 1$*

ET	N_0	N_1	N_3	N_4
0	1030	48200	24700	24000
25	1030	48200	891	19500
50	1030	48200	764	780
75	85	36	68	70
100	85	36	68	70

Table 2.4. *Packets sent with $SEID = 1.5$*

The previous tables show that N_1 is the source because it is transmitting a lot of packets. N_0 is the destination, it is only sending control packets. Nodes N_3 and N_4 are forwarding the packets from N_1 to N_0. If the parameter $SEID$ is increased, the control packet sent by N_0 decreases, because increasing $SEID$ increases the transmission delay between control packets. Using a value $SEID \geq 1$ minimizes the amount of control packets, but the energy information will be less up to date and the algorithm will be less efficient.

For a threshold of $ET = 0$, the sensors N_3 and N_4 forward the same amount of packets, the energy capacity cannot be influenced and the AODV protocol is used. For a threshold of $ET = 25$, the sensor N_4 is favored because it has a great energy capacity. If the threshold is too high $ET \geq 75$, the communication is interrupted, because too much capacity is required.

The following tables show the energy capacity left in the sensors:

ET	N_0	N_1	N_3	N_4
0	9950	9910	9000	9800
25	9960	9920	9070	9810
50	9960	9920	9130	9830
75	9970	9940	9420	9880
100	9970	9940	9420	9880

Table 2.5. *Energy left with $SEID = 0.1$*

The study is focused on the sensor N_3 that has the least energy. For the value $SEID = 0.1$ and $ET = 0$, N_3 has consumed 10% of its energy. This value decreases to 8.5% for $SEID = 0.5$, due to an excess consumption of sending control packets. If $SEID \geq 1$, the algorithm is less efficient. The value $SEID = 0.5$ is chosen for the study.

ET	N_0	N_1	N_3	N_4
0	9960	9920	9150	9830
25	9960	9930	9220	9840
50	9970	9930	9280	9860
75	9970	9950	9450	9890
100	9970	9950	9470	9890

Table 2.6. *Energy left with $SEID = 0.5$*

For the value $SEID = 0.5$ and $ET = 0$, the energy consumption for N_3 is 8.5%, and for $ET = 25$, it decreases to 7.8%. For the value of $ET = 50$, the consumption decreases again, but the number of transmitted packets drops dramatically.

ET	N_0	N_1	N_3	N_4
0	9960	9920	9170	9830
25	9970	9930	9240	9850
50	9970	9930	9300	9860
75	9990	9980	9830	9960
100	9990	9980	9830	9960

Table 2.7. *Energy left with $SEID = 1$*

ET	N_0	N_1	N_3	N_4
0	9960	9920	9180	9840
25	9960	9930	9260	9850
50	9970	9930	9300	9850
75	9990	9990	9900	9980
100	9990	9990	9900	9980

Table 2.8. *Energy left with $SEID = 1.5$*

This routing protocol shows that it is possible to use the mean filter to avoid poor energy regions.

2.4.3. *Routing protocol using deformations*

In this section, a routing protocol using path deformation to minimize the curve energy $E(u)$ defined in equation [2.15] is described. To compute this energy, a rigidity matrix $\mathbf{K}$ and a potential energy P have to be defined.

2.4.3.1. *The rigidity matrix*

A route is a sequence of n sensors, say $N_0, N_1, \ldots N_n$. Between two consecutive sensors, a vector $u^i = \overrightarrow{N_i N_{i+1}}$ can be defined. A path can be represented by a $2n$ vector u with coordinates defined by the coordinates of vectors u^i:

$$u = (u_x^1, u_y^1, \ldots, u_x^n, u_y^n)^T \in \mathbb{R}^{2n}$$

If the sensors are located on a grid, then the coordinates are such that $(u_x^i, u_y^i) \in \{-1, 0, 1\}^2$.

The rigidity matrix $\mathbf{K}$ makes the shape of the route. This matrix is used in the deformation energy:

$$S(u) = \frac{1}{2} u^T \mathbf{K} u$$

The shortest path is a line, and we that this shape will minimize the deformation energy $S(u)$. The matrix $\mathbf{K}$ is constructed to respect this condition:

DEFINITION 2.1.– Let u be a path of length n. Then, the rigidity matrix $\mathbf{K}$ of the path u is defined as a $2n \times 2n$ matrix:

$$\mathbf{K} = \begin{pmatrix} 1 & 0 & 0 & 0 & \ldots & 0 & 0 \\ 0 & 1 & 0 & 0 & \ldots & 0 & 0 \\ 0 & 0 & 2 & 0 & \ldots & 0 & 0 \\ 0 & 0 & 0 & 2 & \ldots & 0 & 0 \\ \vdots & & & & \ddots & & \vdots \\ 0 & 0 & 0 & 0 & \ldots & n & 0 \\ 0 & 0 & 0 & 0 & \ldots & 0 & n \end{pmatrix}$$

The following lemma shows that the deformation energy is lower for a shorter path.

LEMMA 2.1.– Let $S(u)$ be the deformation energy defined by:

$$S(u) = \frac{1}{2} u^T \mathbf{K} u.$$

Let u be a path of length n and w a path of length m with the same source and destination, then the following relation is verified:

$$n < m \Rightarrow S(u) \leq S(w).$$

PROOF.– We have:

$$S(u) = \frac{1}{2} \sum_{i=1}^{n} i\left((u_x^i)^2 + (u_y^i)^2\right)$$

$$S(w) = \frac{1}{2} \sum_{i=1}^{m} i\left((w_x^i)^2 + (w_y^i)^2\right)$$

The proof is done by induction on $m \geq 2$.

For $m = 2$, the sensors are located on a grid and $u_\bullet^1 \in \{-1, 0, 1\}$. $S(u)$ is equal to 1 or $1/2$. If $S(u) = 1/2$, then u is a vertical or horizontal vector, w is a deformation of u and has the same source and destination. This implies that w is made of a diagonal and a horizontal or vertical vector. Then, we have $S(w) = 3/2$. If $S(u) = 1$, then u is a diagonal vector and w is made of a vertical and horizontal vector. Then, we have $S(w) = 3/2$ and $n < m = 2 \Rightarrow S(n) \leq S(w)$.

The relation $n < m \Rightarrow S(u) \leq S(w)$ is verified by induction. The path w is decomposed by a path w' of length m and a path of length 1, and the relation is applied for $n = 1$ and m.

2.4.3.2. Potential energy

The potential energy on a route is defined as the sum of the energies available on the different sensors on the path.

DEFINITION 2.2.– Let u be a path of length n. Each sensor $S^i(S_x^i, S_y^i), 0 \leq i \leq n$ on this path has $I(S_x^i, S_y^i)$ energy available in its battery. Then, the *potential energy* of the path u, say $P(u)$, is defined by:

$$P(u) = -\sum_{i=1}^{n} I\left(u_x^0 + \sum_{k=1}^{i} u_x^k, u_y^0 + \sum_{k=1}^{i} u_y^k\right)$$

where (u_x^0, u_y^0) are the coordinates of the source sensor and (u_x^k, u_y^k) is the vector from sensor S^{k-1} to sensor S^k.

2.4.3.3. *Route deformations*

A route is a path of sensors. A deformation of a route is a path with the same source and destination and with a length smaller or equal to the length of the initial route.

DEFINITION 2.3.– Suppose that the sensors are located on a grid. Let u be a route of source sensor $S_0(u_x^0, u_y^0)$, destination sensor $S_n(u_x^n, u_y^n)$ and length n. A route w of length m is a deformation of u if and only if the following conditions are verified:

$$
\begin{cases}
u_x^0 = w_x^0 \\[6pt]
u_y^0 = w_y^0 \\[6pt]
\displaystyle\sum_{i=1}^{n} u_x^i = \sum_{j=1}^{m} w_x^j \\[6pt]
\displaystyle\sum_{i=1}^{n} u_y^i = \sum_{j=1}^{m} w_y^j \\[6pt]
m \le n
\end{cases}
$$

By direct computation, it is possible to classify the deformations of lengths 1 and 2:

LEMMA 2.2.– Let u be a route of length 2 and w a deformation of u of length 1. Then, this deformation is of type D_1:

$$
\begin{pmatrix} u_x^1 \\ 0 \\ 0 \\ u_y^2 \end{pmatrix} \rightarrow \begin{pmatrix} u_x^1 \\ u_y^2 \end{pmatrix}
\qquad
\begin{pmatrix} 0 \\ u_y^1 \\ u_x^2 \\ 0 \end{pmatrix} \rightarrow \begin{pmatrix} u_x^2 \\ u_y^1 \end{pmatrix}
$$

$$
\begin{pmatrix} u_x^1 \\ u_y^1 \\ -u_x^1 \\ 0 \end{pmatrix} \rightarrow \begin{pmatrix} 0 \\ u_y^1 \end{pmatrix}
\qquad
\begin{pmatrix} u_x^1 \\ 0 \\ -u_x^1 \\ u_y^2 \end{pmatrix} \rightarrow \begin{pmatrix} 0 \\ u_y^2 \end{pmatrix}
$$

$$\begin{pmatrix} u_x^1 \\ u_y^1 \\ 0 \\ -u_y^1 \end{pmatrix} \rightarrow \begin{pmatrix} u_x^1 \\ 0 \end{pmatrix} \quad \begin{pmatrix} 0 \\ u_y^1 \\ u_x^2 \\ -u_y^1 \end{pmatrix} \rightarrow \begin{pmatrix} u_x^2 \\ 0 \end{pmatrix}$$

Figure 2.13 represents a deformation D_1.

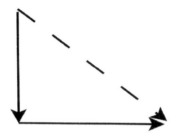

Figure 2.13. *Deformation D_1*

LEMMA 2.3.– Let u be a deformation of length 2 and w a deformation of u of length 2. Then, this deformation is of type D_2:

$$\begin{pmatrix} u_x^1 \\ u_y^1 \\ u_x^1 \\ 0 \end{pmatrix} \rightarrow \begin{pmatrix} u_x^1 \\ 0 \\ u_x^1 \\ u_y^1 \end{pmatrix} \quad \begin{pmatrix} u_x^1 \\ 0 \\ u_x^1 \\ u_y^2 \end{pmatrix} \rightarrow \begin{pmatrix} u_x^1 \\ u_y^2 \\ u_x^1 \\ 0 \end{pmatrix}$$

$$\begin{pmatrix} u_x^1 \\ u_y^1 \\ 0 \\ u_y^1 \end{pmatrix} \rightarrow \begin{pmatrix} 0 \\ u_y^1 \\ u_x^1 \\ u_y^1 \end{pmatrix} \quad \begin{pmatrix} u_y^1 \\ 0 \\ u_x^1 \\ u_y^2 \end{pmatrix} \rightarrow \begin{pmatrix} u_x^1 \\ u_y^2 \\ u_y^1 \\ 0 \end{pmatrix}$$

Figure 2.14 represents a deformation of type D_2.

Lemma 2.4 establishes a relation between route deformation and its energy.

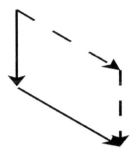

Figure 2.14. *Deformation D_2*

LEMMA 2.4.– Let u be a route of length 2 from sensor S_{n-1} to sensor S_{n+1} via S_n, and $D_1(u)$ its deformation of length 1, then:

– The energy of the deformed route $S(D_1(u))$ is less as $S(u)$:

$$S(D_1(u)) \leq S(u).$$

– The potential energies $D_1(u)$ and u are related by:

$$P(D_1(u)) = P(u) - I(S_n).$$

Let D_2 be a deformation of type 2:

– If $u_\bullet^0 \neq 0$, then:

$$S(D_2(u)) \geq S(u).$$

– If there exists $u_\bullet^0 = 0$, then:

$$S(D_2(u)) \leq S(u).$$

– The potential energies of $D_2(u)$ and u are related by:

$$P(D_2(u)) = P(u) + I(S_n) - I(S'_n),$$

where $D_2(u)$ is the path (S_{n-1}, S'_n, S_{n+1}).

The deformations of type D_1 decrease the energy of a path and will always be applied. The deformations of type D_2 decrease the energy of a path if the potential energy of the intermediate sensor S_n decreases. This deformation will be applied if needed.

2.4.3.4. Routing protocol

The routing protocol will apply deformations to a route to minimize the curve energy E. It first applies D_1 deformations and D_2 deformations are applied after if they reduce the curve energy E. This is done by each sensor in an infinite loop. Because the deformations D_1 and D_2 are applied to the path of length 2, a sensor only needs to know its neighbors and the neighbors of its neighbors.

A deformation D_x acts on three successive sensors. The route u is split into disjoint sets of three sensors. The deformation is applied on each triple set. Figure 2.16 shows this application on a different family of sets. The deformation D_2 is only applied if the energy $E(S_{n-1}, S_n, S_{n+1})$ decreases.

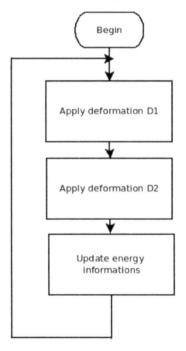

Figure 2.15. *Infinite loop*

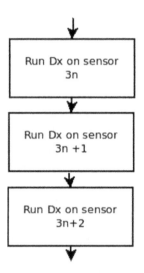

Figure 2.16. *Application of deformations* D_x

2.4.3.5. *Simulations*

The Scilab software is used for simulations. The network is represented by a 300×300 pixel gray level image. The initial energy is set by the uniform distribution on the interval $[e_{min}, 1]$ with $e_{min} = 10/256$. The minimal energy required to route packets is $1/256$. The source sensor is located at the corner $(0, 0)$ and the destination at the corner $(300, 300)$. The initial route is a vertical and horizontal path going through the corner $(300, 0)$. Figure 2.17 shows the initial route like a black line.

Figure 2.17. *Initial route*

Figure 2.18 shows the route after 2000 iterations. The route is the border between the black region of the image and the gray level region. A diagonal starts to be drawn, it will be the shortest path from source to destination.

Figure 2.18. *Route after 2000 iterations*

Figure 2.19 shows the route after 10000 iterations. The diagonal at the bottom right represents the optimal route, but it is extended by a curve. This curve is dependent on the energy capacity of the sensors. The isolated white pixels in the dark region represent sensors with high energy capacity.

Figure 2.19. *Route after 10000 iterations*

Figure 2.20 shows the route after 24000 iterations. The route is the diagonal, the shortest path from source to destination. The dark region under the diagonal represents these sensors with low energy. Their energy was consumed by the packet forwarding process.

Figure 2.20. *Route after 24000 iterations*

The next simulation is done with an energy hole in the middle to show how the route will be constructed to avoid this hole. Figure 2.21 represents the initial energy with the energy hole. The initial route is shown in black like before.

Figure 2.21. *Initial image with hole*

Figure 2.22 shows the state after 2000 iterations. This image is similar to Figure 2.18. Note that the hole has no influence at this state because the route does not reach the energy hole.

Figure 2.22. *Image with hole at 2000 steps*

Figure 2.23 is taken after 10000 iterations. In this image, the route has reached the hole. The route can be split into three distinct paths. The first one is a diagonal starting from the bottom right corner of the image. Then, it follows the lower edge of the hole. After this, it continues to follow a diagonal starting from the lower left corner of the hole.

Figure 2.24 shows the state after 24000 steps. Starting at the left corner of the hole, the route reaches the left border of the image. After this the route is still a vertical line to the source.

2.5. Conclusion

In this chapter, devoted to routing in wireless sensor networks, we proposed an analogy between a grayscale image and the distribution of energy in sensor networks. This analogy is made by associating the position of a sensor to the corresponding pixel in an image and the level of energy of its battery to the gray level of the pixel. Thus, the bright areas contain high-energy sensors, and the dark areas contain the low-energy sensors.

Figure 2.23. *Image with hole at 10000 steps*

Figure 2.24. *Image with hole at 24000 steps*

To obtain a route for sensors with high energy capacity, a path must be constructed in the bright areas of the image. Here, image processing algorithms help us. Routing algorithms have been proposed using filters based on convolution matrices. The detection of contours with the Sobel matrices allowed us to modify the AODV routing protocol to obtain routing that favors paths via energy-rich sensors. Another algorithm using the mean filter was presented. Finally, a protocol based on route deformations was introduced.

These analogy networks of wireless sensors and images open a new approach for the creation of routing algorithms. Although the first application is the search for paths minimizing energy consumption, other parameters can be taken into account such as throughput, congestion or network equity. It is also possible to imagine optimized multiparameter routings using color images.

2.6. Bibliography

[ABD 12] ABDELSALAM H.S., OLARIU S., "Toward adaptive sleep schedules for balancing energy consumption in wireless sensor networks", *IEEE Transactions on Computers*, vol. 61, pp. 1443–1458, 2012.

[ALA 11] ALAM M.M., BERDER O., MENARD D. *et al.*, "Traffic-aware adaptive wake-up-interval for preamble sampling MAC protocols of WSN", *International Workshop on Cross Layer Design*, pp. 1–5, 2011.

[AMM 10] AMMAR I., AWAN I., MIN G., "An improved S-MAC protocol based on parallel transmission for wireless sensor networks", *International Conference on Network-Based Information Systems*, pp. 48–54, 2010.

[BAB 03] BABIC Z., MANDIC D., "An efficient noise renioval and edge preserving convolution filter", *TELSIKS 2003: Proceedings of the 6th International Conference on Telecommunications in Modern Satellite, Cable and Broadcasting Service*, IEEE, pp. 538–541, Belgrade, Serbia, 1–3 October 2003.

[BAS 02] BASU M., "Gaussian-based edge-detection methods-a survey", *IEEE Transactions on Systems, Man, and Cybernetics, Part C: Applications and Reviews*, vol. 32, no. 3, pp. 252–260, 2002.

[CHE 12] CHEN G., ZHANG X., YU J. *et al.*, "An Improved LEACH algorithm based on heterogeneous energy of nodes in wireless sensor networks", *International Conference on Computing, Measurement, Control and Sensor Network*, pp. 101–104, 2012.

[CHU 09] CHUNJIANG Z., YONG D., "A modified Sobel edge detection using Dempster-Shafer theory", in QIU P.H., YIU C., ZHANG H. *et al.* (eds), *Proceedings of the 2009 2nd International Congress on Image and Signal Processing*, IEEE, Tianjin, China, 17–19 October 2009.

[DOO 12] DOOHAN N.V., MISHRA D.K., TOKEKAR S., "Energy Aided Shortest Path Routing Protocol (EASPRP) for highly data centric wireless sensor networks", *International Conference on Intelligent Systems, Modelling and Simulation*, IEEE Computer Society, pp. 652–656, 2012.

[FAL 12] FALCON R., LIU H., NAYAK A. *et al.*, "Controlled straight mobility and energy-aware routing in robotic wireless sensor networks", *International Conference on Distributed Computing in Sensor Systems and Workshops*, IEEE Computer Society, pp. 150–157, 2012.

[GIL 09] GILG M., YOUSEF Y., LORENZ P., "Using image processing algorithms for energy efficient routing algorithm in sensor networks", *Future Computing, Service Computation, Cognitive, Adaptive, Content, Patterns, Computation World*, IEEE Computer Society, pp. 132–136, 2009.

[GOY 12] GOYAL D., TRIPATHY M.R., "Routing protocols in wireless sensor networks: a survey", *Advanced Computing & Communication Technologies, International Conference on Anvanced Computing and Communication Technologies*, IEEE Computer Society, pp. 474–480, 2012.

[HAN 11] HANEEF M., DENG Z., "Comparative analysis of classical routing protocol LEACH and its updated variants that improved network life time by addressing shortcomings in wireless sensor network", *Sixth International Conference on Mobile Ad-hoc and Sensor Networks*, IEEE Computer Society, pp. 361–363, 2011.

[HU 11] HU J., MA Z., SUN C., "Energy-efficient MAC protocol designed for wireless sensor network for IoT", *International Conference on Computational Intelligence and Security*, IEEE Computer Society, pp. 721–725, 2011.

[HUA 11] HUANG J.-W., HUNG C.-M., YANG K.-C. *et al.*, "Energy-efficient probabilistic target coverage in wireless sensor networks", *IEEE International Conference on Networks*, IEEE Computer Society, pp. 53–58, 2011.

[HUA 12] HUANG Y.-F., WANG L.-M., TAN T.-H. *et al.*, "Performance of a novel energy-efficient data relaying in wireless sensor networks", *International Symposium on Computer, Consumer and Control*, IEEE Computer Society, pp. 793–796, 2012.

[JIA 09] JIANLAI W., CHUNLING Y., CHAO S., "A novel algorithm for edge detection of remote sensing image based on CNN and PSO", in QIU P.H., YIU C., ZHANG H. *et al.* (eds), *Proceedings of the 2009 2nd International Congress on Image and Signal Processing*, IEEE, Tianjin, China, 17–19 October 2009.

[JIN 09] JIN-YU Z., YAN C., XIAN-XIANG H., "Edge detection of images based on improved Sobel operator and genetic algorithms", in MIN Y., ZHAO X.M., ZHANG Z.J.J. *et al.* (eds), *Proceedings of 2009 International Conference on Image Analysis and Signal Processing*, IEEE, Taizhou, China, April 11–12, 2009.

[KAS 12] KASI M.K., HINZE A., LEGG C. *et al.*, "SEPSen: semantic event processing at the sensor nodes for energy efficient wireless sensor networks", in *Proceedings of the 6th ACM International Conference on Distributed Event-Based Systems*, ACM, pp. 119–122, New York, USA, 2012.

[LAH 12] LAHMAR K., CHEOUR R., ABID M., "Wireless sensor networks: trends, power consumption and simulators", *Asia International Conference on Modelling & Simulation*, IEEE Computer Society, pp. 200–204, 2012.

[LI 12] LI Z., SHEN H., "A Kautz-based real-time and energy-efficient wireless sensor and actuator network", *IEEE 32nd International Conference on Distributed Computing Systems*, IEEE Computer Society, pp. 62–71, 2012.

[MAD 12] MADHU S., DAHIYA A., DAHIYA B., "Energy efficient data transfer in secure wireless sensor networks", *International Conference on Advanced Computing & Communication Technologies*, IEEE Computer Society, pp. 495–499, 2012.

[MAZ 12] MAZINANI S.M., NADERI A., SETOODEFAR M. *et al.*, "An energy-efficient real-time routing protocol for differentiated data in wireless sensor networks", *IEEE International Conference on Engineering of Complex Computer Systems*, IEEE Computer Society, pp. 302–307, 2012.

[MCI 96] MCINERNEY T., TERZOPOULOS D., "Deformable models in medical image analysis: a survey", *Medical Image Analysis*, vol. 1, no. 2, pp. 91–108, 1996.

[PAT 12] PATHAK A., ZAHEERUDDIN, LOBIYAL D. *et al.*, "Improvement of lifetime of wireless sensor network by jointly effort of exponential node distribution and mixed routing", *International Conference on Communication Systems and Network Technologies*, IEEE Computer Society, pp. 316–319, 2012.

[SAR 11] SARANGI S., KAR S., "Genetic algorithm based mobility aware clustering for energy efficient routing in wireless sensor networks", *IEEE International Conference on Networks*, IEEE Computer Society, pp. 1–6, 2011.

[SHA 12] SHARMA M., SHARMA K., "An energy efficient extended LEACH (EEE LEACH)", *International Conference on Communication Systems and Network Technologies*, IEEE Computer Society, pp. 377–382, 2012.

[SHI 12] SHISONG X., XIANGLING Z., FENG Z. *et al.*, "Energy-based cluster partition method in wireless sensor networks", *International Conference on Computational and Information Sciences*, IEEE Computer Society, pp. 912–915, 2012.

[TOY 12] Toyoda S.-N., Sato F., "Energy-effective clustering algorithm based on adjacent nodes and residual electric power in wireless sensor networks", *International Conference on Advanced Information Networking and Applications Workshops*, IEEE Computer Society, pp. 601–606, 2012.

[WAN 12a] Wang L., Li L., "A combined algorithm routing protocol based on energy for wireless sensor network", *International Conference on Computer Science and Electronics Engineering*, IEEE Computer Society, vol. 1, pp. 224–228, 2012.

[WAN 12b] Wang X., Wang J., Lu K. *et al.*, "GKAR: a novel geographic K-anycast routing for wireless sensor networks", *IEEE Transactions on Parallel and Distributed Systems*, vol. 99, IEEE Computer Society, 2012.

[YI 11] Yi X.-S., Jiang P.-J., Wang X.-W. *et al.*, "Survey of energy-saving protocols in wireless sensor networks", *International Conference on Robot, Vision and Signal Processing*, IEEE Computer Society, pp. 208–211, 2011.

[YOU 10a] Yousef Y., Gilg M., Lorenz P., "Using convolution filters for energy efficient routing algorithm in sensor networks", *International Journal on Advances in Intelligent Systems*, vol. 3, nos. 1–2, pp. 150–161, 2010.

[YOU 10b] Yousef Y., Gilg M., Lorenz P., "Using matrix convolutions and clustering for energy efficient routing algorithm in sensor networks", *Sixth Advanced International Conference on Telecommunications*, IEEE Computer Society, pp. 275–279, 2010.

[ZHA 12] Zhang Q., Qu W., "An energy efficient clustering approach in wireless sensor networks", *International Conference on Computer Science and Electronics Engineering*, IEEE Computer Society, vol. 1, pp. 541–544, 2012.

[ZHU 12] Zhu X.Z., Li Y.F., "Simulation of coverage problem research in wireless sensor networks based on energy saving", *International Conference on Computer Science and Electronics Engineering*, IEEE Computer Society, vol. 1, pp. 270–273, 2012.

Routing and Data Diffusion in Vehicular Ad Hoc Networks

3.1. Introduction

Data delivery is a crucial task in vehicular networks since current applications require the cooperation of each and every vehicle. Regarding the interaction between the driver and the vehicle, as defined by the National Highway Traffic Safety Administration [NHT 16], a fully autonomous car performs a given driving task under any variety of conditions. Therefore, all decisions taken by the system become crucial, simply because no actual human hand is on the "steering wheel" of the driving system. Information provided by sensors make it possible to construct a local vision of the vehicle; however, the construction of a global vision of the actual situation at hand requires the exchange of information and data. This task is ensured by data delivery services, with the aim of being efficient and reliable. Security aspects must also be taken into account in order to disseminate, receive and process reliable information and data.

The term vehicular ad hoc network (VANET) usually refers to all wireless vehicular networks. Considered as a subset of mobile ad hoc networks (MANET) by [COR 99], VANETs share common features: a dynamic topology due the mobility of nodes, a limited available bandwidth and limited security since the communication medium is shared by all stations. However, vehicles do not have limited energy capacity, because they have an alternator.

Chapter written by Frédéric DROUHIN and Sébastien BINDEL.

Regarding these features, a data delivery service has to ensure reliable communication despite the mobility of nodes, minimize the bandwidth consumption and secure communication. In this chapter, data delivery service in VANET is investigated through three aspects: (i) how to select a destination, (ii) how to route data to a destination and (iii) how to secure communication. The selection of destination is performed through a transmission method, which can reach one or several destinations. In praxis, a special kind of address is assigned to identify nodes such as the unicast, broadcast, multicast or anycast addresses. Data routing is ensured by routing protocols, which determines the path between two non-adjacent vehicles. The computation of the best path relies on information provided by a metric which assesses the "cost" of each path to reach a desired destination. Communication security provides additional services according to the application requirement. Two kinds of security must be considered: passive attacks and active attacks. Within passive attacks, only monitoring tasks are performed, unlike in active attacks wherein an action is performed by a hacker.

The remainder of this chapter is organized as follows. Section 3.2 describes the context and challenges related to each considered aspect of the data delivery service. In section 3.3, routing protocols related to vehicular networks are detailed. In section 3.4, security aspects are detailed. Section 3.5 closes this chapter and provides outlook.

3.2. Background and challenges

The deployment of routing and security solutions requires compliance with the characteristics and standards of vehicular networks. Regarding the features of VANET, the high dynamic of the topology and the uncertain and random density of vehicles have a significant impact on the connectivity to the network and the delivery delay. Information provided by on-board sensors gives local information, such as the position, useful for commutation. Communications in vehicular networks rely on three architectures. The first one is *the Vehicle-to-Vehicle ad hoc network (V2V)* where vehicles communicate directly to each other and form a fully distributed network. The second one is *the Vehicle-to-Infrastructure network (V2I)* wherein vehicles communicate only with the roadside infrastructure via RoadSite Units (RSU) and form a centralized network. The last one is the *hybrid architecture* combining both the V2V and V2I infrastructures. A vehicle can communicate either in a single hop or multi-hop fashion. The design of routing and security solutions needs to take into account the network architecture, the communication standards defining the protocol stack, and the signal propagation to understand the disturbances generated by the environment.

3.2.1. *Communication standard*

The dedicated short-range communication (DSRC) system has been specifically used for vehicular communications. It is a short/medium-range technology that operates at the 5.9 GHz band that has been widely standardized. The most investigated standard is certainly the one designed by the Institute of Electrical and Electronics Engineers (IEEE). It includes two standards, the IEEE 802.11p and the Wireless Access in Vehicular Environments (WAVE).

The standard IEEE 802.11p was introduced in 2004 as an amendment of the IEEE 802.11 in order to address vehicular communication. It describes the requirements of the physical and data link layers and is part of the WAVE architecture dedicated to intelligent transport systems (ITS). The physical amendment of 802.11p is similar to the IEEE 802.11a, both work in the range of 5 GHz but have a different bandwidth, 20 MHz for 802.11a and 10 MHz for 802.11p. Table 3.1 lists the remaining differences. Assuming a theoretical communication range up to 1000 m (V2V and V2I), [GAL 06] have shown in praxis that the maximum range in line of sight (LOS) is 880 m and in non-line of sight (NLOS) between 58 m and 230 m.

Parameters	IEEE 802.11a	IEEE 802.11p	Changes
Bit rate(Mb/s)	6, 9, 12, 18, 24, 36, 48, 54	3, 4.5, 6, 9, 12, 18, 24, 27	Half
Code rate	BPSK, QPSK, 6QAM, 64QAM	BPSK, QPSK, 16QAM, 64QAM	No change
Number of subcarriers	1/2, 2/3, 3/4	1/2, 2/3, 3/4	No change
Symbol duration	52	52	No change
Guard time	$4\mu s$	$8\mu s$	Double
FFT period	$0.8\mu s$	$1.6\mu s$	Double
Preamble duration	$3.2\mu s$	$6.4\mu s$	Double
Subcarrier spacing	0.3125 MHz	0.15625	Half

Table 3.1. *IEEE 802.11a and IEEE 802.11p parameters*

The aim of WAVE architecture is to give wireless access in a vehicular environment. The standard defined two stacks, one dedicated to the data plan, Figure 3.1, the other dedicated to the management plan, a resource manager and a security service. Regarding the data plane, WAVE includes IEEE 802.11p amendment to define physical and the lower layer of the data link. In WAVE, the channel is split in two, one half dedicated to signalization, called the Control Channel (CCH) and the other dedicated to information transmission over IP, called the Service Channel (SCH). This functionality is detailed in the IEEE 1609.4. The IEEE 1609.3 standard defined network services including Logical Link Control (LLC), IP and transport layers and the management plane layer. The resource manager defined in the IEEE 1609.1 standard runs at the application layer and is destined to manage services provided by the applications. Security services defined in the IEEE 1609.2 define security mechanisms for applications and manage messages to guarantee confidentiality, authenticity, integrity and anonymity.

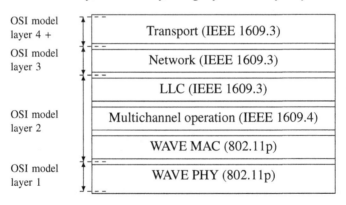

Figure 3.1. *WAVE communication stack: data plane*

3.2.2. *Signal disturbance*

In wireless networks, the environment plays a significant role in network performances because it disturbs the signal propagation. This feature has to be considered in the design of routing protocols in order to be suitable for vehicular networks. An electromagnetic wave is made up of an electric field (E) and a magnetic field (B), oscillating at the same frequency and spreading in the same direction, as depicted in Figure 3.2.

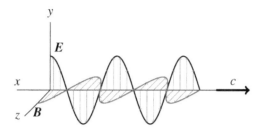

Figure 3.2. *Illustration of an electromagnetic wave*

The distance between two oscillations is called the wavelength and denoted λ (*m*). Let *c* be the speed of light $(3.10^8 \text{ m.s}^{-1})$ and *f* the frequency (5.9 GHz for VANET), then the wavelength is computed as follows:

$$\lambda = \frac{c}{f} = \frac{3 \times 10^8}{5.9 \times 10^9} \approx 0.0508m \cdot \qquad [3.1]$$

Four main effects responsible for the signal disturbance can be distinguished. First, path loss, which represents the attenuation of the signal between the emitter and the receiver. Second, large-scale shadowing describes a fading occurring on a large scale. Third, small-scale fading occurring on a small scale. Fourth, the Doppler effect which is the change of the wavelength between an emitter and a receiver in motion. Figure 3.3 depicts the effect of path loss, shadowing and the multi-path effect versus the distance.

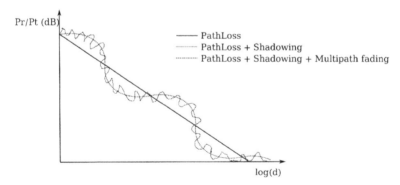

Figure 3.3. *Path loss, shadowing and multi-path effects versus the distance*

3.2.2.1. *Path loss*

In a perfect environment, electromagnetic waves are only affected by the frequency and the distance between the transmitter and the receiver. In such an environment, no obstacles are present between stations, and it is denoted as *free space*. Let G_t be the transmission gain, P_t the power of the transmitted signal, and W the power density at a distance d, which is computed as follows according to [PAR 00] :

$$W = \frac{G_t P_t}{4\pi d^2}. \qquad [3.2]$$

From the relationship between W and P_r the received signal power, equation [3.3] can be formulated as follows :

$$P_r = WA_r = W\frac{\lambda^2 G_r}{4\pi}, \qquad [3.3]$$

with A_r the effective aperture of the received antenna, λ the wavelength and G_r the reception gain. From equations [3.2] and [3.3], the [FRI 46] equation determines the signal attenuation in a free space environment:

$$\frac{P_r}{P_t} = G_t G_r \left(\frac{c}{4\pi f d}\right)^2 \qquad [3.4]$$

The Friis equation describes a vanilla environment and can be considered suitable for describing a signal propagation in far field environments. Regarding the ground environment and the position of the antennas, the propagation loss model described by Friis can be improved by taking into account the signal reflection on the ground. This model is called a two ray ground and describes the line of sight component and the multi-path component (ground reflection) of the received signal. According to [RAP 01], for a very large distance d and a perfect polarization and reflection, the calculation of the received signal power P_r can be formulated according to the equation [3.5].

$$P_r = P_t G_t G_r \frac{h_t^2 h_r^2}{d^4 L}, \qquad [3.5]$$

with h_t and h_r the height of the transmitter and receiver antenna and L the system loss and fixed at 1 according to [3.5]. A common strategy adopted by network simulators is to use the two-ray ground model, equation [3.5], when the distance d is larger than a cross-over distance d_c and a free-space model, equation [3.4], in the other case:

$$P_r = \begin{cases} \frac{P_t G_t G_r c^2}{(4\pi f d)^2} & d \le d_c, \\ P_t G_t G_r \frac{h_t^2 h_r^2}{d^4 L} & d > d_c, \end{cases}$$

[3.6]

with $d_c = 4\pi \frac{h_t h_r}{\lambda}$.

Friis and the two-ray ground model are not suitable for describing dense environments, such as urban or building environments. A common model adopted for describing such environments is the log-distance model, where the path loss L at a distance d is expressed as follows:

$$L = L_0 + 10\alpha \log_{10} \frac{d}{d_0}$$

[3.7]

with L_0 the reference path loss value based on measurement made at distance d_0. A close version of this model applies three consecutive log-distance models with different α coefficients according to the distance between the emitter and the receiver.

3.2.2.2. Large-scale shadowing

Large-scale shadowing, also called local mean attenuation, occurs due to the obstruction of the signal when it meets an obstacle such as a building, a truck or an hill. Confirmed by empirical studies in both indoor and outdoor environments, the most common model used is log-normal shadowing. It takes into account the path loss based on a reference distance d_0 and applies a log-normal shadowing in order to compute the power reception $\overline{P_r(d)}$ as follows:

$$\frac{P_r(d_0)}{\overline{P_r(d)}} = (\frac{d}{d_0})^\beta + X_{\sigma_{dB}},$$

[3.8]

with a typical value for $\beta \in [2.7, 5]$ and X_σ is a log-normal distribution with a standard deviation $\sigma_{dB} \in [4, 12] dB$ in an outdoor environment.

3.2.2.3. *Small-scale fading*

Small-scale fading refers to the effect of the multi-path of the wave. It occurs when the interaction between the signal and the obstacle produces a split of the current signal with a different speed and strength. Such effects can be modeled by some statistical laws. The Rayleigh model describes an environment with several multi-paths that have the same strength. Such models are used to describe environments where the signal is highly disturbed. The amplitude of the received signal z can be described by the Rayleigh distribution defined by equation [3.9].

$$P_z(z) = \frac{z}{\sigma^2} \exp\left(\frac{z^2}{2\sigma^2}\right), \tag{3.9}$$

with $2\sigma^2$ the root mean square of the received signal and the assumption that each path is uniformly distributed $[-\pi, \pi]$. An environment wherein the strength of a path is higher than others can be described with a Rice distribution. The amplitude of the received signal, z, can be described with the Rice distribution defined in equation [3.10].

$$P_z(z) = \frac{z}{\sigma^2} \exp\left(-\frac{z^2 + A^2}{2\sigma^2}\right) I_0\left(\frac{z.A}{2\sigma^2}\right), \tag{3.10}$$

where I_0 is the Bessel function (zero order) and A the amplitude of the predominant path. The strength of the predominant path and others ratio is determined by $K = \frac{A^2}{2\sigma^2}$ describing the fading degree. Regarding K, the Rice distribution can model a Rayleigh if $K = 0$ and a Gaussian distribution when $K \to \infty$.

3.2.2.4. *Doppler*

The motion of an emitter and a receiver produces a frequency shift of the incoming electromagnetic waves. It results in an offset in the carrier frequency as depicted in Figure 3.4, where two observers A and B are looking for a vehicle going to B direction. A typical scenario is the sound emitted by a moving car with an observer behind it and an observer in front of the car. For A observer, the sound frequency is higher than the frequency measured at the car, whereas the frequency at B is lower than that measured at the car.

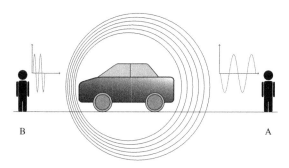

Figure 3.4. *Illustration of the Doppler effect*

The resulting frequency f_r observed when a source emits at a frequency of f is computed as follows:

$$f_r = f\frac{c + v_r}{c - v_s},$$ [3.11]

where c is the celerity of the wave, v_s the velocity of the source and v_r the velocity of the observer. Regarding the motion of the source, v_s and v_r are positive if the receiver is moving towards the source and the receiver is moving away from the receiver, respectively. Under three assumptions, the antenna of the receiver is omnidirectional, the radio wave is propagated horizontally and the angle of receiving radio waves is uniformly distributed, the Doppler effect can be described by a Jakes model, wherein the normalized Jakes Doppler power spectrum is given by:

$$S(f) = \frac{1}{\pi f_d \sqrt{1 - (f/f_d)^2}}, |f| \leqslant f_d,$$ [3.12]

where f_d is the maximum Doppler frequency.

3.2.2.5. *A note on praxis measurements*

In a theoretical manner, a link between two entities is considered as enabled or lost. In praxis, such a vision is too simplistic. As depicted in Figure 3.6, the signal affected by a path loss, shadowing and a fast-fading effect cannot be easily considered as enabled or lost. [ZAM 07] have demonstrated with their channel model that a link is characterized by three phases (Figure 3.5). First, the connected region where the link has a high packet reception ratio. Second, the transitional region, wherein the link is considered as unreliable and the packet ratio is fluctuated. Third, the disconnected region, where the link has a low packet reception ratio.

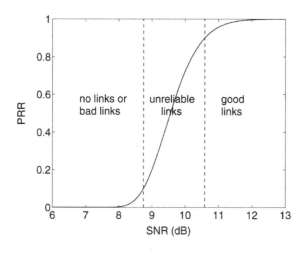

Figure 3.5. *The three regions of the link quality*

In Figure 3.6, the signal affected only by a Friss model shows a disconnection at 2000 m. With realistic propagation models, the good region of the link extends to 700 m, at which point the transitional region is then extended to around 1500 m, and consequently at the end, the list is then considered to be disconnected.

It is important for performance evaluations to test data delivery solutions under realistic propagation conditions. That is why [ABB 15] have designed a realistic propagation model from real measurements. Its effectiveness was investigated and confirmed by [HIL 17].

3.3. Routing

This section presents routing protocols and their related mechanisms. In VANETs, vehicles have different velocities and are driven in different environments contributing to a decrease in the connectivity of the network. In such a situation, the design of routing protocols need to take into account these features. The first part of this section presents metrics used by routing protocols to assess the local links and thus the routing path. The second part details dedicated routing protocols for both V2V and for V2I infrastructures.

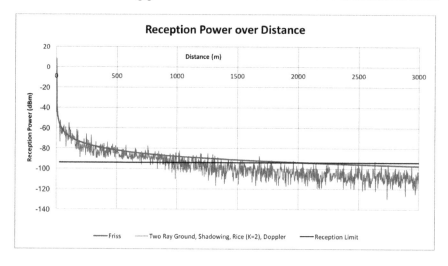

Figure 3.6. *Attenuation, shadowing, fast fading effect: power reception over distance with two ray ground, shadowing, Rice (K = 2) and Doppler model*

3.3.1. *Neighbor selection*

Path selection relies on information provided by metric, given a "cost" to reach a neighbor. It can be related to the number of hops, the link quality, the bandwidth, the latency, etc. The computation of the path cost is then related to the cost of each local link composing it and four types of metrics can be distinguished according to [WAN 99]. The first one includes additive metrics, where the cost of the path is the sum of all costs of local links. The second one regroups multiplicative metrics, where the cost of the path is the multiplication of all costs of local links. The third group is the concave metric, where the cost of the path is the minimum cost of local links. The last group includes convex metrics, wherein the cost of the path is the maximum cost of local links. Owing to the high velocity of vehicles in VANET, most of the dedicated routing protocols use a combination of metrics in order to have an accurate assessment of the cost of the link. This section starts with a description of the most used metrics in VANET and details current implemented solutions.

3.3.1.1. *Hop count metric*

The traditional approach takes into account the number of hops to reach a destination. The ultimate benefit of such a technique is that it is user-friendly. This hop count number is currently coded on 8 bytes and incremented by one at each retransmission. Widely used by routing protocols dedicated to wired networks, such an approach becomes inefficient in wireless networks as pointed out by [DEC 03b]. Authors showed through a wireless sensor test, that the shortest paths often have fewer capacities than others and have opened a discussion on the link quality.

3.3.1.2. *Link quality estimators*

Link quality estimators (LQE) have been well designed for wireless sensor networks, and are also used in vehicular networks. [BAC 12] have classified these estimators into two categories. The first category, hardware-based, includes all estimators performing an assessment from information available at the physical layer. The second, software-based, regroups the rest of LQE running at upper layers, either on MAC or the IP layer.

3.3.1.2.1. Hardware LQE

Hardware-based estimators perform an assessment from information available at the physical layer. Assessments are provided by the receiver hardware without any additional computation costs, and are performed only by the receiver at each frame reception. In addition, the computed link quality can be assessed though the current traffic on the wireless channel without the need for any periodical broadcasts. In order to evaluate the reliability of an estimator, a good fitting with the Packet Reception Ratio (PRR) is required.

The first estimator, the received signal strength indicator (RSSI), gives the signal strength of the received packet. The signal-to-noise ratio (SNR) gives the difference between the pure received signal strength and the noise floor. The link quality indicator (LQI) only available in the IEEE 802.15 networks provides a link quality assessment based on the height symbols of the received packet. A second generation of hardware-based LQE has been designed based on detailed information on the decoding process related to the DSSS (direct-sequence spread spectrum) used in IEEE 802.15.4 networks. [HEI 12] have designed an estimator called CEPS, which relies on chip errors on the payload symbols to assess the PRR. As demonstrated by the authors, the correlation between the chip errors and the PRR can be approximated by a linear fit. Later, [SPU 13] suggested the BLITZ estimator, an improvement on the CEPS, by also considering chip errors related to the preamble. This improvement allows analysis of packet synchronization errors in order to obtain faster and more accurate information on the link quality.

As depicted in Figure 3.7(a), the correlation between the RSSI and the PRR is not easily deductible.

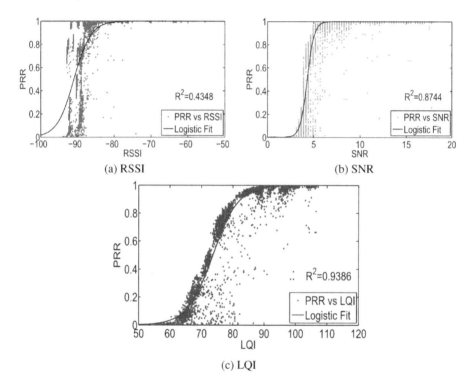

(a) RSSI

(b) SNR

(c) LQI

Figure 3.7. *PRR as a function of RSSI, SNR, LQI in WSN for 160 hours of data though 72 links got by [LIU 14]*

However, [SRI 06] have demonstrated that over a RSSI threshold, the PRR is consistently high. This observation can be confirmed on the Figure 3.7(a), where the PRR is above 0.9 when the RSSI is below −85 dBm. Regarding the SNR, it provides a better correlation with the PRR than the RSSI; however, a simple observation is not sufficient to deduct the corresponding PRR. The same observation can be made for the LQI, even if the coefficient correlation is the highest. As a result, regardless of the estimator, a single reading is insufficient to determine the PRR. With CEPS and BLITZ, authors try to solve this deficiency. The two estimators present a better correlation with the PRR, and Blitz can provide an assessment as soon as a preamble is detected, even if the frame reception failed. The major drawback of such solutions is their implementation, since they require information from the decoding process only available in the chipset radio. That is why all experiments have been performed with a software radio.

3.3.1.2.2. Software LQE

Software-based LQE performs an assessment from information available either at the MAC layer or at the network layer. In this context, estimators can only detect the good reception or the loss of a packet. Unlike hardware estimators, their assessments correlate directly with the PRR. Three kinds of estimators can be considered: (i) PRR-based, (ii) RNP-based and (iii) score-based estimators.

3.3.1.2.3. PRR based

PRR-based estimators are the simplest link quality estimators. Their computation is based on the ratio between received packets and expected packets. Considered as a reference for hardware-based, they can only assess the quality of the downlink part. The efficiency of such an estimator is the most challenging, since it relies on the time window size used by the moving average. For wireless sensor networks, a good practice is to fix the window size according to the channel coherence, since the velocity of nodes can be considered as near zero. Let c be the celerity of light $(m.s^{-1})$, fc the carrier frequency in (Hz) and v the velocity $(m.s^{-1})$; the channel coherence Tc is then computed as follows:

$$T_c = \sqrt{\frac{9}{16\pi f_d^2}} \approx \frac{1}{4f_d},$$

$$\text{with } f_d = \frac{vf_c}{c}. \qquad\qquad [3.13]$$

Regarding the velocity of vehicles, the PRR cannot be computed from the time channel coherence. There is no silver bullet to compute the PRR regardless of the link lifetime, but some solutions try to solve this issue. [BIN 15] have suggested the use of a dynamic window size and [WOO 03] propose the use of the window mean with exponentially weighted moving average (WMEWMA).

3.3.1.2.4. RNP based

RNP-based estimators assess both sides of a link (downlink and uplink sides). A bidirectional communication is then required, which is why such estimators can detect unidirectional links. [CER 05] have demonstrated the absence of a relationship between the required number of packets (RNP) and the reception rate (RR). From this observation, they have designed an estimator maintained at the sender side and based on the required number of packet transmissions (RNP) before a successful reception. The RNP estimator runs at the link layer and requires the use of an automated repeat request (ARQ) to repeat the transmission of unicast packets until it is received. [DEC 03a] have designed the expected transmission count (ETX) estimator, which is maintained at the receiver side. Based on the computation, several ratios, d_f, are related to the PRR and, d_r, to the acknowledgment reception ratio (ARR). [FON 07] have designed the four-bit metric to be easily used by routing protocols and provide four bits of information related to each layer. The "white" bit set from the physical layer indicates whether the medium quality is high. The "black" bit provided by the local link layer indicates if an acknowledgment has been received for a transmitted packet. The network layer provides two pieces of information through the "pin" and "compare" bits and are used for the neighbor table replacement policy.

3.3.1.2.5. Score based

This category regroups all estimators providing a score rather than a value related to a specific phenomena such as the packet reception or the number of retransmissions. [BAC 10] have developed the fuzzy link quality estimator (F-LQE) mixing four indicators, the smoothed reception ratio, the link stability factor, the link asymmetry level and the channel average signal-to-noise ratio. F-LQE combines these information by using the fuzzy logic rule, in order to provide a single metric for the routing protocol. In [REN 11], authors designed an estimator called holistic packet statistics (HoPS), which provides information on the static and dynamic behavior of the link through four estimators. One is dedicated to providing a short-term assessment of the packet success rate, and another is dedicated to providing a long-term assessment. From the two, the absolute deviation and the trend are computed in order to assess the stability of the link. The F-ETX metric developed by [BIN 16] assesses both the quality and the state of the link. Concerning the link quality, F-ETX provides a short-term and a long-term assessment in order to compare links with a transient quality. Beside, F-ETX determines the link stability and transient and persistent unidirectionality phases. The main drawback of score-based metrics concerns the integration of information return by the metric in the routing process. There is no silver bullet to combine information to get an only metric value, authors of F-LQE use fuzzy logic; meanwhile, [REN 11] use empirical methods. The approach suggested by [BIN 16] was to integrate each piece of information in the routing process, e.g. if a link is declared unstable, then another link will be looked for.

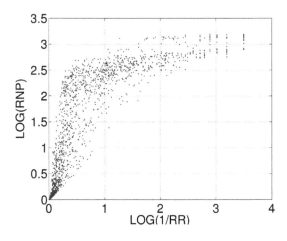

Figure 3.8. *RNP as a function of 1/RR*

3.3.2. *Routing protocols*

Routing protocols dedicated to MANET have been widely investigated in the past. Considered as a subset of MANET, some routing protocols from MANET can also be used in VANET. However, the high velocity of vehicles and their specific motion have been taken into account in the design of specific routing protocols for VANET. A first survey made by [LI 07] introduced a classification based on the following criteria: *ad hoc*, position-based, cluster-based, broadcast and geocast-based. Even if an author has the merit of giving a classification and introducing concepts, they have become out of date due to recent research advances. [LEE 10] and [SHA 14] investigate in depth routing solutions. The most recent classification is given by [SHA 14] and is based on the VANET architecture, V2V and V2I. Routing protocols dedicated to the V2V architecture can be classified into six categories: (i) topology-based, (ii) position-based, (iii) cluster-based, (iv) geocast-based, (v) multicast-based and (vi) broadcast-based. Figure 3.9 depicts the different delivery scheme used by unicast, broadcast, multicast and geocast protocols. Concerning routing protocols for V2I, two groups are considered: (i) for static infrastructure and (ii) for mobile infrastructure.

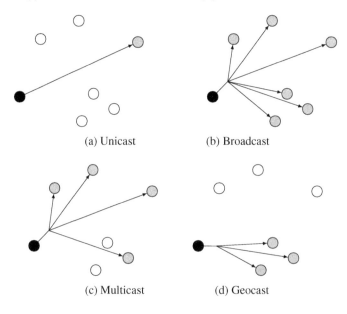

(a) Unicast   (b) Broadcast

(c) Multicast   (d) Geocast

Figure 3.9. *Type of delivery scheme*

3.3.2.1. *V2V routing protocols*

Routing protocols designed for a V2V architecture exploit the distributed feature of such a network. The most beneficial aspects are the loss of tolerance and its scalability; moreover, it is often hard to maintain stability, which therefore requires several approaches that have been investigated and detailed in the following paragraphs.

3.3.2.1.1. Topology based

Topology-based protocols have been inherited from MANET and rely on topology information to construct a routing path. These protocols are classified into three categories: proactive, reactive and hybrid. *Proactive protocols*, also called table-driven, maintain a periodic routing path discovery. This strategy has the benefit of providing nodes with a fresh vision of the topology, but it requires a significant bandwidth, not available for data. The most popular protocol is certainly the optimized link state routing (OLSR) protocol, but there are others such as: destination-sequenced distance vector (DSDV), fisheye state routing (FSR), global state routing protocol (GSRP), wireless routing protocol (WRP) and topology dissemination based on reverse-path forwarding routing (TBRPF). *Reactive protocols*, also called on-demand routing protocols, start the routing discovery process when data need to be transmitted to a desired destination. These protocols have the benefit of having a reduced bandwidth consumption, unlike proactive protocols; however, since the routing process is triggered only by data needed to be transmitted, the delay related to the routing path discovery is then added. The most popular protocol is *ad hoc* on-demand distance vector (AODV) but there are others: temporally ordered routing algorithm (TORA), prediction-based AODV (PRAODV) and dynamic source routing (DSR). The last category includes protocols that combine both a proactive and a reactive routing scheme. Most protocols define a zone wherein a proactive scheme is used and a reactive scheme to route data inter-zone. The main idea of this approach is to profit from the benefits of proactive and reactive schemes and avoid drawbacks. The best known protocols are zone routing protocol (ZRP) and hybrid *ad hoc* routing protocol (HARP).

3.3.2.1.2. Position based

Position-based protocols rely on the geographic node's location acquired with a specific device such as a Global Positioning System (GPS). This information is used for making routing decisions. Such protocols can be classified into three categories: non-delay tolerant, delay tolerant and hybrid. *Non-delay tolerant* protocols aim to transmit data to a destination as soon as possible. A typical routing strategy is to forward data to the node closest to the destination. This category includes Greedy Perimeter Stateless Routing (GPSR), Geographic Source Routing (GSR), etc. *Delay tolerant* tolerant protocols adopt another approach by keeping data packets when a direct end-to-end path is not available. The biggest drawback of such a solution is the possible important end-to-end delay. These protocols are efficient when the node density is low. The best known protocols are scalable knowledge-based vehicular routing (SKVR), vehicle-assisted data delivery (VADD) and geographical opportunistic (GeOpps). The last category, *Hybrid position-based protocols*, combines both a delay and a non-delay routing scheme. The aim is to profit from the network connectivity to choose the best routing algorithm. When a path exists, the non-tolerant algorithm is used, but when the path is broken and no route exists, the tolerant delay algorithm is triggered. This category includes the GeoDTN+NAV protocol.

3.3.2.1.3. Cluster based

Cluster-based protocols form a cluster in order to reduce the number of broadcasting dates by reducing the diffusion domain. A cluster is formed by a cluster head that is responsible for its management. Communications inside a cluster are performed directly between vehicles. For intercommunications, vehicles have to forward data to the cluster head. The creation of a cluster can be static or dynamic; however, the management of dynamic clusters is more difficult. The most common cluster-based routing protocols are cluster-based routing (CBR), location routing algorithm with cluster-based flooding (LORA-CBF), clustering for open IVC network (COIN) and traffic infrastructure-based cluster routing protocol with handoff (TIBCRPH).

3.3.2.1.4. Geocast based

Close to the position-based protocols, geocast-based protocols use a multicast routing to deliver a message to all vehicles situated in a geographical area. The aim of such protocols is to deliver a packet from a source node to all nodes in the same geographical area. Geocast routing can be considered as a multicast service where the network is split according to the desired geographical region. This category includes the following protocols: Inter-Vehicle Geocast (IVG), Cached Geocast Routing (CGR) and Mobicast.

3.3.2.1.5. Multicast based

Unlike geocast-based protocols where packets are delivered to all nodes situated in a region, multicast-based routing protocols maintain a structure such as a tree or a mesh structure to define destination nodes. Tree-based routing protocols attempt to maintain a multicast routing tree to transfer from a source to a group of destination nodes. The main drawback of such an approach is that the tree needs to be rebuilt when the topology is volatile, as a result of which routing service is continuously disrupted. Multicast-based routing protocols rely on a tree structure, which include the multicast ad hoc on demand (MAODV) protocol, adaptative demand-driven multicast routing protocol (ADMR) and multicast with ant-colony optimization for VANETS based on MAODV (MAV-AODV) protocol. Mesh-based multicast routing protocols maintain a connected component of the network containing receivers of a group. Two protocols can be considered as mesh-based multicast routing protocols: On-demand multicast routing protocol (ODMRP) and destination-driven on-demand multicast routing protocol (D-ODMRP).

3.3.2.1.6. Broadcast based

The last category includes broadcast-based protocols that flood routing messages over the network. This strategy enhances the probability reception of a message to a destination, but with a high bandwidth cost. Suitable for scattered networks, they become less efficient when the density increases. Broadcast-based protocols include BROADCOMM, urban multi-hop broadcast (UMB) and distributed vehicular broadcast (DV-CAST).

3.3.2.2. *V2I routing protocols*

The high mobility of vehicles does not allow V2V protocols to be efficient in all situations. In this context the use of V2V routing protocols only is not sufficient. That is why some routing protocols exploit both the vehicles and the infrastructure. Regarding the infrastructures, they can be considered as static or dynamic.

3.3.2.2.1. Static infrastructure

Assuming fixed RSU linked to a backbone, some routing protocols need a uniform distribution of RSU; meanwhile, others suppose a placement only at intersection. For example, the static node-assisted adaptive (SADV) routing protocol can only consider RSU placed at intersections, but the roadside-aided routing (RAR) require RSU to be placed at each extremity of a geographical area.

3.3.2.2.2. Dynamic infrastructure

RSU tends to minimize the end-to-end delay, the main problem is over the covered area. To solve this issue, some research suggests replacing fixed RSU with mobile vehicles to obtain a mobile infrastructure such as for the mobile infrastructure-based VANET routing (MIBR), the mobile gateway routing protocol (MGRP) and the prediction-based routing (PBR) protocols.

3.4. VANET security

The VANET offers a multitude of services ranging from accident prevention, multimedia and Internet access. These different uses are strongly linked to computer security. Moreover, the VANET protocol stack references the open system interconnection (OSI) network model, and so, inherits from its vulnerability. Indeed, by taking a simple example with a vehicle (*malicious vehicle*), which broadcasts alert messages, it is easy to cause congestion or even accidents. This simple example shows the need to integrate IT security into VANETs.

Safety goes beyond accident prevention even and remains a priority. Risks related to the interception of data that may compromise both vehicle (e.g. owner data of the manufacturer, vehicle location) and driver (privacy data such as home location). Through these examples, security is at the heart of the VANET issues such as the recent work of [ABB 16] on the controller area network (CAN) bus safety. In addition, like all connected devices, vehicles can be used as *botnets* to relay attacks of the type Deny of Service (like MIRAI botnet) and consequently cause congestion of network traffic.

The attack objectives vary and depend on the target of a hacker. They may want to alter the proper functioning of a system, destabilize a company or even a country, steal data, trade secrets, private data as mentioned above, in order to use or resell it and of course to serve as an emblem of a given hacker's dubious skills. These attacks can be carried out by individuals, a set of constituted and coordinated individuals, rival companies, foreign governments, but also the government of a country (e.g. population supervision). These attacks do not necessarily seek to be destructive but can alter the proper functioning of the networks and thus cause varying amounts of damage. Depending on the type of attack, company employees (seeking revenge) can also participate in these attacks and have a much greater impact.

The VANET security protocols must guarantee the important notions of security: authentication, non-repudiation, integrity but also the private data of the manufacturer and the driver and his passengers. It concerns vehicles, RSU and both V2V and V2I communication.

The first section describes the security requirements in VANETs generalized to the IoT, the second section gives the various attacks in terms of passive attacks and active attacks and the last section discusses VANET security solutions.

3.4.1. *Security requirements in VANET*

[KER 16] and [XIA 05] describe the protection against different attacks using various requirements in VANET security set-up. These requirements are:

– *Authenticity*: data authentification ensures that a message is trustworthy and sent by a legitimate and authorized vehicle.

– *Integrity*: data should not be altered or modified by an unauthorized third party. Modifications may be intentional or due to faulty sensors.

– *Non-repudiation*: is the mechanism to associate a transaction with the emitter. The emitter cannot deny that the message was sent by itself.

– *Availability*: communication channel should be available to allow vehicles to send information and other vehicles to receive.

– *Access control*: a transaction sent should be reliable and secure and altered messages removed by an authority.

– *Confidentiality*: when exchanging data, the confidentiality of data should be guaranteed. In VANETs, vehicles are also anonymous from the point of view of other vehicles and from RSUs. On the other hand, they must be recognized by a trusted authority.

All these requirements are mandatory in order to ensure security in VANET.

3.4.2. *VANET security threats*

In this section, major attacks are described. In a wireless network, passive listening is all the easier as the air medium is difficult to control. Passive attacks do not change the operation of the system but seek to collect information about the system. As part of the VANET, the hacker will seek to collect a set of information about the vehicle (theft of industrial secrets) or data from the private life of the driver. Indeed, the attacks also concern the passengers since the VANETs also include playful aspects and, with the extension of the WiFi in vehicles as a new service, passengers are also sensitive to these attacks. Moreover, the layers between the on-board WiFi and the VANET communication (or even CAN bus) should be studied in order to be sure that intercommunication is not possible. Active attacks cause the attacked system to malfunction. Hackers seek to disrupt the system in order to render it inoperative or no longer able to perform the service for which it is made. Both passive and active attacks pass by the analysis and monitoring of the traffic (i.e. messages exchanged). Even if the analysis and the monitoring are passive, it is the basis of any attack. It consists of intercepting all of the traffic and then detecting a security breach or collected data. Hackers may *monitor and analyze a network* to collect all information or use *brute force* attacks by generating a large number of consecutive values usually breaking encryption keys. The brute force attack can be time- and resource-consuming.

Different attacks concern security and confidential threats and concern both on-board units (OBU) and RSU. Note that the RSU should be actively protected since they may be used as authority or at least manage the communication authority. RSU also manage traceability that coexists with confidentiality. Attacks do not only refer to a given hacker's targets but can be a consequence of faulty devices or captors (e.g. wrong temperature, wrong node speed or wrong location). Faulty devices may cause wrong interpretations similar to attacks by hackers (deny-of-service, delay in delivery, etc.), and finally, spread faulty information which may cause congestion and safety issues like masquerading attacks.

[MEJ 14] provides a list of different possible attacks:

– *Availability*:

- *Denial Of Service* (DoS): one or several nodes (distributed denial of service - DDoS) flood a network by sending continuous (dummy) messages that overload a network and make it unusable or at least reduces network efficiency. This method can also target a single and specific node.

- *Jamming*: at the level of propagation, it consists of transmitting on the same frequency ranges as VANET.

- *Black hole attack*: since a node can route and forward messages to other nodes, it can drop all the traffic and discard all packets. This node may be related to a sink node. A gray hole attack does not drop all packets but selects information type (e.g. safety) or randomly drops some information.

- *Malware*: like any computer-based system, VANET is sensitive to viruses, worms, trojans, spyware, adware, rootkits, ransomware, etc. The differences with the previous cited attacks is that VANET need to install a third-party software, but VANET are still affected in case of snooping attacks (spyware, adware), it can also be used for data modification (viruses) or as a relay in an attack dedicated to VANETs or even for a more global attack on the Internet (ransomware, rootskits, trojans, worms).

– *Integrity*:

- *Replay attack*: consists of replaying the original message emitted by an authenticated and authorized vehicle. This kind of attack affects the network (flooding) and resources on vehicles (CPU, memory, etc.), and, of course, alters VANET service. This attack does not focus on stopping the operation of VANET like a denial of service attack.

- *Data modification attack*: an active attack, it is based on interception of exchanged data. This data can be modified and deleted in order to alter the comprehension of the message and to prevent information arriving at the receivers, e.g. in the case of an accident or traffic congestion.

– *Authenticity and identification*:

- *Replay attack and masquerading attack*: an already described in the previous paragraph (Integrity).

- *GPS spoofing*: one or several nodes (malicious or infected nodes) send fake locations which affect geographical protocols or service applications based on GPS.

- *Timing attack*: consists of delaying messages, especially safety messages, or prevents information arriving on time to receivers (expired information).

- *Repudiation*: a node denies a message that has been sent by itself which requires sending the message a second time (time- and resource-consuming).

- *Sybil attack*: similar to a botnet, hackers launch attacks using controlled nodes (malicious or infected nodes) to relay other attacks (e.g. replay, timing, DoS attacks). It can also be used to propagate an attack on VANET or Internet nodes.

- *Masquerading attack*: can be used in an impersonation attack where the authorized and authenticated vehicle provides a valid identity to the attacker. The node can turn into a malicious node and send fake alerts or malicious messages: betrayal attack.

– *Confidentiality*:

- *Traffic analysis*: monitor and analyze network to collect information and find security breaches. Once enough data are collected, they might be modified, altered or stolen.

- *Eavesdropping*: e.g. man in the middle, essentially intercepting communication.

- *Snooping attack*: mainly concerns privacy data both from industrial and driver/passengers information. In this case, it concerns information from drivers' licenses and car information to Global Positioning System (GPS) position of home/work place, etc.

Table 3.2 summarizes attacks and attacked services based on requirements in section 3.4.1.

Type of attacks	Service attacked
GPS Spoofing	Authenticity, Non-repudiation
Masquerading attack	Authenticity
Replay attack	Authenticity, Availability, Integrity
Repudiation	Authenticity, Non-repudiation
Sybil attack	Authenticity, Non-repudiation
Data modification attack	Authenticity, Availability, Integrity
Black/Gray hole attack	Availability
Jamming	Availability
DoS/DDoS	Availability
Timing attack	Availability
Traffic Analysis	Confidentiality, Privacy
Brute Force Attacks	Confidentiality, Privacy
Eavesdropping	Confidentiality, Privacy
Snooping attack	Privacy
Illegal Tracking	Privacy

Table 3.2. *Security threats and solution category*

3.4.3. *VANET security mechanisms: IEEE 1609.2-2016 standard*

The previous section discussed the different possible attacks. Some research on solutions against the different attacks listed, [MEJ 14] and [SUN 10], provide an interesting survey. In this section, IEEE 1609.2-2016 standard is discussed.

IEEE 1609.2-2016 (IEEE Standard for Wireless Access in Vehicular Environments and Security Services for Applications and Management Messages) proposes a standard with the following definition: "this standard defines secure message formats and processing for use by Wireless Access in Vehicular Environments (WAVE) devices, including methods to secure WAVE management messages and methods to secure application messages. It also describes administrative functions necessary to support the core security functions". This standard is used in IEEE 1609.3-2016 for WAVE Service Announcement security and in SAE J2945/1-201603, On-Board System Requirements for V2V Safety Communications, for Basic Safety Message security. This standard provides the following requirements:

– Secure protocol data unit (PDU) format for signed data and encrypted data: it provides payload, hash of external payload, provider service ID to indicate permissions with optional fields (generation time, expiry time, generation location, security management), reference to signing certificate and signature.

– Certificate format for signing PDUs applications with pseudonymous (no identification of sender) and identifier: certificate contains permissions (service-specific permissions) and a provider service ID together with a signed secured PDU.

– Certificate authorities (CA): all messages are signed by a certificate which is provided by a certificate authority in cascade with at least one certificate in the list known and trusted by a receiver.

– Certificate revocation list (CRL) format that allows revoking or invalidating for different reasons (e.g. private key compromised, change in certificate).

– Peer-to-peer certificate distribution to allow new certificates: this requirement is mandatory and added to the list of certificates with always the feature that one certificate is known in the list. Receiver should be able to build a cascade of certificates to a trusted and identified certificate.

To be a valid message, the receiver has to check that the signed secure PDU has verified that none of the certificates have been revoked, one certificate in the list is trusted, the signature is verified, the payload is consistent with the provider service ID and permissions and the message is relevant (recent, not expired, not a replay). The data are encrypted with symmetric key with a persistent public key. Concerning the exchange of certificate, it is based on asymmetric cryptography (public and private keys) that requires the establishment of a public key infrastructure (PKI). PKI provides several security services with a trusted CA with confidentially, authenticity, integrity and non-repudiation.

This standard is still in development and different research projects (e.g. Crash Avoidance Metrics Partnership) are providing input for its development.

3.4.3.1. *Summary*

Similar to propagation models, security protocols impact network performance and computing capabilities. There exist several cryptographic approaches to be applied in VANET, including public key programs to distribute session keys for message encryption, authentication schemes and random traffic patterns against traffic analysis. Constraints such as privacy (e.g. position detection) must be consistent with the traceability of messages required by law enforcement authorities. In addition, the constraint in message delivery time should not be impacted by cryptography. IEEE 1609.2-2016 provides requirements for security on several attacks. These different solutions must also be coupled with stand-alone vehicle systems such as Lidar, cameras and other sensors to ensure better security and reliability, particularly in a safety context.

3.5. Conclusion

The aim of this chapter was to describe routing and security solutions for vehicular ad hoc networks. The background details communication standards and signal disturbances. This information must be taken into account in the design of routing security protocols. Standards define protocol format messages and how a solution can be implemented in the network stack. Furthermore, signal disturbance investigation gives an overview on challenges met by routing protocols which have to ensure efficient data delivery services. A detailed investigation has been conducted into routing solutions dedicated to vehicular networks. It first details a theoretical overview of routing algorithms, then it describes the current metrics used for the node selection process and, finally, practical routing protocols are detailed and compared. Finally, a survey on security aspects has been proposed and it shows urgent challenges in such networks.

There is no silver bullet, and current routing protocols are not efficient in all situations. The dynamic topology in VANET is currently being studied through dynamic graphs, but no algorithm can guarantee no packet loss and a time delay boundary. As a result, the discussion is still open. Concerning security, this aspect cannot be ignored in the design of solutions dedicated to vehicular networks. A consortium regrouping the vehicle industry (Mercedes-Benz, BMW, Audi, Opel, Ford, Boss Continental, etc.) and public institutions are working on safety solutions such as the Safe Intelligent Mobility project.

This conclusion finishes by mentioning 5G developed by [GEN 17] and specifically the release of 14 (in development) that proposes an alternative for 802.11p (V2I and V2V) with point-to-point communication (device to device) and includes a wide range of road users (e.g. pedestrians, bicyclist, etc.) especially in a safety schema.

3.6. Bibliography

[ABB 15] Abbas T., Sjöberg K., Karedal J. *et al.*, "A measurement based shadow fading model for vehicle-to-vehicle network simulations", *International Journal of Antennas and Propagation*, vol. 2015, p. 12, 2015.

[ABB 16] Abbott-McCune S., Shay L.A., "Techniques in hacking and simulating a modem automotive controller area network", *2016 IEEE International Carnahan Conference on Security Technology (ICCST)*, pp. 1–7, October 2016.

[BAC 10] Baccour N., Koubâa A., Youssef H. *et al.*, "F-LQE: a fuzzy link quality estimator for wireless sensor networks", *Proceedings of the 7th European Conference on Wireless Sensor Networks*, EWSN'10, Heidelberg, pp. 240–255, 2010.

[BAC 12] Baccour N., Koubâa A., Mottola L. *et al.*, "Radio link quality estimation in wireless sensor networks: a survey", *ACM Transactions on Sensor Networks*, ACM, vol. 8, no. 4, pp. 34:1–34:33, September 2012.

[BIN 15] Bindel S., Chaumette S., Hilt B., *F-ETX: An Enhancement of ETX Metric for Wireless Mobile Networks*, Springer International Publishing, 2015.

[BIN 16] Bindel S., Chaumette S., Hilt B., "F-ETX: a predictive link state estimator for mobile networks", *ICST Transactions on Mobile Communications Applications*, vol. 2, no. 7, p. e3, 2016.

[CER 05] CERPA A., WONG J.L., POTKONJAK M. *et al.*, "Temporal properties of low power wireless links: modeling and implications on multi-hop routing", *Proceedings of the 6th ACM International Symposium on Mobile Ad Hoc Networking and Computing*, MobiHoc '05, ACM, New York, USA, pp. 414–425, 2005.

[COR 99] CORSON S., MACKET J., "Mobile ad hoc networking (MANET); routing protocol performance issues and evaluation considerations", *RFC 2501*, January 1999.

[DEC 03a] DE COUTO D.S.J., AGUAYO D., BICKET J. *et al.*, "A high-throughput path metric for multi-hop wireless routing", *Proceedings of the 9th Annual International Conference on Mobile Computing and Networking*, ACM, MobiCom '03, New York, USA, pp. 134–146, 2003.

[DEC 03b] DE COUTO D.S.J., AGUAYO D., CHAMBERS B.A. *et al.*, "Performance of multihop wireless networks: shortest path is not enough", *SIGCOMM Computer Communication Review*, vol. 33, no. 1, pp. 83–88, ACM, January 2003.

[FON 07] FONSECA R., GNAWALI O., JAMIESON K. *et al.*, "Four bit wireless link estimation", *Proceedings of the Sixth Workshop on Hot Topics in Networks (HotNets VI)*, pp. 1–14, 2007.

[FRI 46] FRIIS H., "A note on a simple transmission formula", *Proceedings of the IRE*, vol. 34, no. 5, pp. 254–256, May 1946.

[GAL 06] GALLAGHER B., AKALSUKA H., SUZUKI H., "Wireless communications for vehicle safety: radio link performance and wireless connectivity methods", *IEEE Vehicular Technology Magazine*, vol. 1, no. 4, pp. 4–24, December 2006.

[GEN 17] 3RD GENERATION PARTNERSHIP PROJECT (3GPP), "5G – Release 14", available at: http://www.3gpp.org/release-14, 2017.

[HEI 12] HEINZER P., LENDERS V., LEGENDRE F., "Fast and accurate packet delivery estimation based on DSSS chip errors", *2012 Proceedings IEEE INFOCOM*, pp. 2916–2920, March 2012.

[HIL 17] HILT B., BERBINEAU M., VINEL A. *et al.*, *Networking Simulation for Intelligent Transportation Systems: High Mobile Wireless Nodes*, ISTE Ltd, London and John Wiley & Sons, New York, June 2017.

[KER 16] KERRACHE C.A., CALAFATE C.T., CANO J.C. *et al.*, "Trust management for vehicular networks: an adversary-oriented overview", *IEEE Access*, vol. 4, pp. 9293–9307, 2016.

[LEE 10] LEE K.C., LEE U., GERLA M., "Survey of routing protocols in vehicular ad hoc networks", *Advances in Vehicular Ad-hoc Networks: Developments and Challenges*, pp. 149–170, 2010.

[LI 07] LI F., WANG Y., "Routing in vehicular ad hoc networks: a survey", *IEEE Vehicular Technology Magazine*, vol. 2, no. 2, pp. 12–22, June 2007.

[LIU 14] LIU T., CERPA A.E., "Data-driven link quality prediction using link features", *ACM Transactions on Sensor Networks*, vol. 10, no. 2, pp. 37:1–37:35, ACM, January 2014.

[MEJ 14] MEJRI M.N., BEN-OTHMAN J., HAMDI M., "Survey on VANET security challenges and possible cryptographic solutions", *Vehicular Communications*, vol. 1, no. 2, pp. 53–66, 2014.

[NHT 16] NHTSA, Accelerating the Next Revolution In Roadway Safety, pp. 1–113, September 2016.

[PAR 00] PARSONS J., *The Mobile Radio Propagation Channel*, Wiley, 2nd edition, October 2000.

[RAP 01] RAPPAPORT T., *Wireless Communications: Principles and Practice*, 2nd edition, Prentice Hall PTR, Upper Saddle River, USA, 2001.

[REN 11] RENNER CHRISTIANAND E.S., WEYER C., TURAU V., *Prediction Accuracy of Link-Quality Estimators*, Springer Berlin Heidelberg, Berlin, Heidelberg, 2011.

[SHA 14] SHAREF B.T., ALSAQOUR R.A., ISMAIL M., "Vehicular communication ad hoc routing protocols: a survey", *Journal of Network and Computer Applications*, vol. 40, pp. 363–396, 2014.

[SPU 13] SPUHLER M., LENDERS V., GIUSTINIANO D., *BLITZ: Wireless Link Quality Estimation in the Dark*, Springer Berlin Heidelberg, Berlin, Heidelberg, 2013.

[SRI 06] SRINIVASAN K., LEVIS P., "RSSI is under appreciated", *Proceedings of the Third Workshop on Embedded Networked Sensors (EmNets)*, 2006.

[SUN 10] SUN J., ZHANG C., ZHANG Y. *et al.*, "An identity-based security system for user privacy in vehicular ad hoc networks", *IEEE Transactions on Parallel and Distributed Systems*, vol. 21, no. 9, pp. 1227–1239, September 2010.

[WAN 99] WANG Z., "On the complexity of quality of service routing", *Information Processing Letters*, vol. 69, no. 3, pp. 111–114, 1999.

[WOO 03] WOO A., CULLER D., Evaluation of efficient link reliability estimators for low-power wireless networks, Report no. UCB/CSD-03-1270, EECS Department, University of California, Berkeley, 2003.

[XIA 05] XIAODONG L.R.L., *Vehicular Ad Hoc Network Security and Privacy*, Wiley-IEEE Press, New York, 2005.

[ZAM 07] ZAMALLOA M.Z.N., KRISHNAMACHARI B., "An analysis of unreliability and asymmetry in low-power wireless links", *ACM Transactions on Sensor Networks*, vol. 3, no. 2, ACM, June 2007.

New "Graphiton" Model: A Computational Discrete Space, Self-Encoded as Trivalent Spin Networks

4.1. Introduction

In this chapter, we present a new type of model particularly suited to solving physical problems in non-Euclidean spaces. These spaces may mix the concepts of proximity (such as geographic information systems), topological notions (such as social networks), information concepts and algebraic concepts (such as the invariance of certain local symmetries).

Our model was developed to provide a framework for digital physics. It is a conceptual breakthrough in the definition of a space based on set theory. We can compute the gravitational field from its metrics. Here, we specify its formalism in information theory, opening up applications to networks [ASC 08, ASC 09a, ASC 09b, ASC 10].

This model is a complex system on several levels, where the fundamental level is that of a trivalent graph.

The second level is the encoding of information in the topological graph structure.

The third is the encoding of local symmetries and algebraic properties.

Chapter written by Smain FEMMAM and Raymond ASCHHEIM.

The fourth is the emergence of a metric, a dimensionality and a curvature of the assembly of subgraphs, as supernodes in a crystalline type network.

It is possible to work at every level of this model by maintaining a good coherence. We optimize the simplicity of the structure at a fundamental level, while allowing the emergence of great complexity in the modeled system.

This family of models is designed to provide a framework for a new theory of everything, unifying, in theoretical physics, the phenomena of quantum physics and our knowledge of the standard model with gravity and the conception of a space according to general relativity. We are dealing with a discrete model instead of continuous one. There is still a debate as to whether the nature of the universe is discrete or continuous. Some even speculate that the reality is the alliance of the two: [CHA 10] *"We examine the hypothesis that space-time is a product of a continuous four-dimensional manifold times a finite space"*.

Others postulate that a discrete (and possibly finite) system is sufficient for the emergence of an apparent continuity, which is an approximation of the discrete. The main result of the present work is the demonstration that a phenomenon hitherto treated as a continuous course, emerges naturally in our discrete model: gravity. As it emerges from a simple graph, we call it *"graphitation"*. A similar play on words has already been used for an approach to quantum physics in a graph, called *"Quantum Graphity"* (see Markopoulou *et al.* [MAR 04] and Konopka *et al.* [KON 08]). However, the approaches are different; physicists have built several discrete models, graphs or networks to implement continuous physical equations, by keeping the current models in discrete format. These approaches are top-down, as they depart from known results to find a consistent underlying structure.

Our approach is bottom-up, starting with a simple model (the level of set theory, the foundation of mathematics), and constructing according to a principle of economy, *"Okham's razor principle"* (*Rursus absque necessitate et utilitate est pluralitas fugienda*) [DE 19, DE 40], a layered model to the more complex reality.

4.2. Graphitation, bottom-up approach

In our bottom-up approach, we start with a trivalent undirected graph.

4.2.1. *Construction*

4.2.1.1. *Definition*

A *graphiton* model is a trivalent graph which checks at least two properties: a quasi-invariance on a large scale (a large number of evenly spaced nodes in the graph are topologically equivalent if we neglect a small number of disturbances) and a characterization of local geometry at small scale by the presence of triangular loops encoding bit 1.

Graphiton, a basic element of a *graphiton* model, is simply a node with three ends, each connected to another *graphiton*.

So, while any totally trivalent graph is a blend of *graphitons*, for it to be a model, it must satisfy the properties of global quasi-invariance and local geometric encoding.

A *graphiton* model can be interpreted in different scales. We will distinguish three scales.

Level 0 is the fundamental level, in which the model is a simple undirected trivalent graph, where any node or connection does not carry additional information.

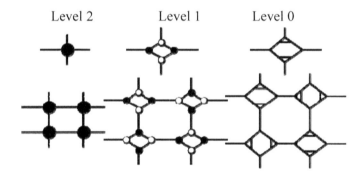

Figure 4.1. *Three-level view of (a) a simple node, (b) a subgraph. At level 2, the graph is not trivalent. At level 1, nodes are added so the graph becomes trivalent, and nodes hold a bit. At level 0, 1-valued nodes are replaced by triangles*

Level 1 is obtained from level 0 by adding a binary property to each node; each node carries a bit, taking the value 0 or value 1.

The equivalence between level 1, informed graph, and level 0, is ensured by the transformation T2 (see Figure 4.3) by Alexander and Briggs [ALE 26], which replaces each node with a value of 1 at level 1 as a triangle consists of three nodes at 0, and does not change the nodes on the value 0 to level 1.

If the graph is *information homogeneous*, if contains as many bit 0 as bit 1 at level 1, whereas its equivalent in level 0 will double its number of nodes (and therefore connections).

Unless explicitly stated otherwise, we always work on information homogeneous graphs.

Our models have a natural encoding and regular, homogeneous information, which can be seen as physically optimum polarization for a certain Hamiltonian, defining a ground state of the system. This system may be disrupted and contain additional information by the inversion of two narrow bits 0 and 1.

We show that this inversion is made only by means of Alexander's T1 transformations (see Figure 4.2), which provide a number of dynamic invariant systems, and indirectly, ownership of causal invariance as discussed by Wolfram [WOL 02]. The T1 transformation is intuitively the simplest, most realistic and most economic possible evolution of a trivalent graph. It applies to a connection between two nodes, and involves two other neighbors of each of these nodes. Of these six nodes forming the face of an H if the connection is in a horizontal view, the T1 transformation rotates H 90 degrees. This can be done on a continuous basis through a transient state in which a tetravalent node 5 is at the center of an X-shaped figure (Figure 4.2 when nodes 5 and 6 are merged).

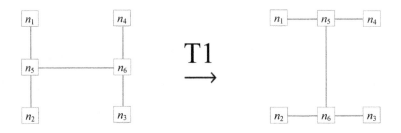

Figure 4.2. *Elementary move: T1 inversion*

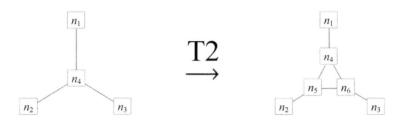

Figure 4.3. *Elementary move: T2 insertion*

At each connection, we can apply two different transformations T1, depending on which of the first neighboring nodes is found connected to a second neighbor node, or the other round.

If the graph is locally planar, one T1 transformation retains planarity, and the other does not, thereby determining an implicit one that will be applied.

4.2.1.2. *Disrupting*

At level 1 (see Figure 4.1), a disturbance is the inversion of values 0 and 1 of two bits carried by two nodes, either neighbor (at distance 0) or narrow (at distance 1, separated by a node):

$$1{-}0 \leftrightarrow 0{-}1$$
$$1{-}1{-}0 \leftrightarrow 0{-}1{-}1, \ 1{-}0{-}0 \leftrightarrow 0{-}0{-}1 \ \text{(Figure 4.4)}.$$

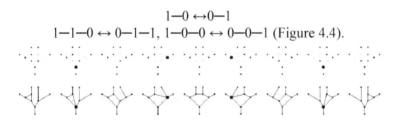

Figure 4.4. *Bit inversion 1–0–0 ↔ 0–0–1 as a sequence of four Alexander T1 moves (the transient four-valent X-state is enhanced by a bigger dot)*

4.2.1.3. *Dimension of the system as a space*

The dimension of the graph depends on the chosen scale. We distinguish three levels:

Ultra-local dimension: around a node, a trivalent graph is of dimension 2 [AMB 04, CAR] and is considered the local restriction of a hexagonal lattice.

Local scale: within a super node, the structure is that of a binary tree graph whose dimension is infinite, because the number of neighbors at a distance r increases as 2^r, and therefore more quickly than r^d for any finite dimensions d.

Global dimension: imposed by the regular system chosen (2 with a grid, 4 with hyperdiamond [ASC 08, ASC 09a, ASC 09b, ASC 10], n-1 with a simplex, etc.).

4.2.2. *Global topology of space*

The global topology is flat if the network is chosen infinite.

It becomes curved if the network is closed, as, for example, in n-dimensional Torus.

Closure can imply a shift if a line does not meet itself after a tour, but a neighbor, like the non-commutative torus.

Closure may reverse the orientation, forming a Moebius strip, a Klein bottle or other non-oriented spaces [LUM 01].

4.2.3. *Lattice regularity*

4.2.3.1. *Regular lattice*

We can apply our model to any regular lattice, like the square lattice, the checkerboard lattice or even lattices in the A, D, E, F families.

We can further investigate applying it to quasi-regular lattices: quasi-crystal [SAD 97] and Penrose tiling [CON 90].

4.2.3.2. *Almost regularity*

We can expand the set of acceptable models to almost regular lattices, defined as regular lattices admitting a number of irregularities $< n^{(d-1)}$ for a graph of n^d nodes and global dimension d.

4.2.3.3. *Local regularity*

As a special case of almost regularity, we can build models which physically correspond to the structure of natural crystals. They admit irregularities on the surfaces of dimension d-1, locally regular but globally isotropic. The space is broken into pieces of irregular crystals which themselves are regular but not parallel to each other.

It will then be useful to study the dynamic behavior of inter-crystalline walls. From a mathematical point of view, almost regularity and local regularity are similar because they induce the same small number of irregularities with respect to the global regularity necessary.

4.2.4. *Intrinsic space*

The fundamental advantage of the *graphiton model* is its creation of intrinsic space. From the outside the system is a graph whose nodes do not coordinate in any space (unlike all previous approaches to discretization of physics). But thanks to the encoding of information in the topology of this graph, the model finds the characteristics necessary for a metric space, beyond the mere internal metrics of the graph, which alone would be insufficient. The *graphiton model* is an information system capable of emulating a space. And a rather rich area because it is Riemannian [PEN 05] and admits a local curvature, it can be Ricci-flat, where, in the absence of disturbance, any node at level 2 is topologically equivalent [ASC 08, ASC 09a, ASC 09b, ASC 10]. Most importantly, local disturbances linearly induce a curvature, then a tilt of geodesics at any point in space, forming a gravitational field. This field is attracted towards disturbances that are at its origin, and is additive; these are necessary conditions for a gravitational field reflecting the effect of mass disruption. Therefore, this space is quite a natural framework for a theory of general relativity. However, if space is locally anisotropic as directions are marked by the regular network, the gravitational induced field remains isotropic and not focused only along the directions of the network. Thus, a movement in the network in response to the gravitational field can follow any direction not subject to the directions of the network.

4.2.5. *Space as a network*

The general idea that the ultimate nature of space is a trivalent graph and some topological characteristics of this graph encode a mass effect, with a result on the curvature and the possibility of calculating a Ricci curvature and the Einstein field equation [PEN 05], was developed by Wolfram [WOL 02].

The fundamental contribution we make to this concept and which enables an effective implementation is to effectively place a layer of information in the graph at level 1, associated with the regular structure at level 2. This layer allows information to move intrinsically within the graph at a selected distance in both directions (two directions for each dimension), and makes it possible to calculate the gravitational field and the Ricci tensor [PEN 05]. In the most random graph models proposed by Wolfram, these calculations were impossible and the notions of Ricci tensor could not be defined clearly. In other discrete models based on regular arrays, the authors postulated the existence of all space, locally Euclidean and continuous extrinsic to their model, in which they plunged their model to a metric and coordinates. What is less well founded and elegant than the concept of Wolfram! We have therefore in the extreme elegance of the model suggested by Wolfram, the only model of intrinsic system that does not need to be plunged into another space to be calculated. But we overcome its lack of metrics by imposing constraints of regularity and topological encoding of the local orientation.

4.3. Mathematical formalism

We give a formal description of our *graphiton model*.

4.3.1. *Definitions*

DEFINITION 4.1.– rank r binary tree:

$$\mathbb{B}_r := \{\underset{k=0}{\overset{2^r-1}{\text{nodes}}}\{n_k\}, \text{links}\{(n_0, n_1) \underset{k=1}{\overset{2^{r-1}-1}{\cup\cup}}(n_k, n_{2k}) \underset{k=1}{\overset{2^{r-1}-1}{\cup\cup}}(n_k, n_{2k+1})\}\} \qquad [4.1]$$

Examples are given in Figure 4.5 for a $\mathbb{B}_3$ graph and its subtree from n_2.

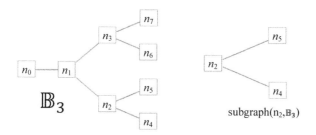

Figure 4.5. *Examples: a $\mathbb{B}_3$ graph (left), its subtree from n_2 (right)*

DEFINITION 4.2.– descendants are a set of nodes descendant from one node:

$$\text{descendants}\left(n_k \text{in} \mathbb{B}_r\right) :=$$

$$\begin{cases} \text{if } k < 2^{r-1} : n_k \cup \text{descendants}\left(n_{2k} \text{in} \mathbb{B}_r\right) \cup \text{descendants}\left(n_{2k+1} \text{in } \mathbb{B}_r\right) & [4.2] \\ \text{else} : n_k \end{cases}$$

DEFINITION 4.3.– subtree is the branch emerging from a node:

$$\text{subtree}\left(n_k \text{in} \mathbb{B}_r\right) := \mathbb{B}_r \cap \text{descendants}\left(n_k \text{in} \mathbb{B}_r\right) \tag{4.3}$$

DEFINITION 4.4.– an operator $\varXi$ glues three binary trees (of the four subtrees) to build a triple-binary trivalent graph:

$$\mathbb{T}_r = \varXi\left(\mathbb{B}_r\right) \text{by} \{ \begin{matrix} \text{removenodes}\{n_3, \text{descendants}\left(n_6\right)\} \text{and their links} \\ \text{addlinks } \left(n_2, n_0\right)\left(n_0, n_5\right)\left(n_1, n_7\right), \text{removelink}\left(n_2, n_5\right) \end{matrix} \tag{4.4}$$

A triple-binary trivalent graph is shown in Figure 4.6 with 24-node module $\mathbb{T}_5 = \varXi\left(\mathbb{B}_5\right)$, rank 5, with 12 leaves.

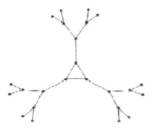

Figure 4.6. *24-node module $T_5 = \varXi\left(B_5\right)$, rank 5, with 12 leaves*

DEFINITION 4.5.– we define the binary decomposition of an integer:

$$\beta(K) := \{b_{i,K}\} \in \{0,1\}^{\lfloor 1 + \frac{\text{Log}K}{\text{Log}2}\rfloor} \mid K = \sum_{i=0}^{\lfloor \frac{\text{Log}K}{\text{Log}2}\rfloor} b_{i,K} 2^i \qquad [4.5]$$

DEFINITION 4.6.– we define the frequency of a quaternion:

$$\omega(x) := \frac{1}{\pi} \text{ArcCos}(\text{Re}(x)) \qquad [4.6]$$

DEFINITION 4.7.– we define the logarithm of a quaternion from its frequency:

$$\log(x) = \log(\mid x \mid) + \pi\omega(x)\frac{\text{Im}(x)}{\mid \text{Im}(x) \mid} \qquad [4.7]$$

DEFINITION 4.8.– we define a family of r couples of a frequency and an imaginary direction:

$$\Omega_r := \{\{\omega_k, u_k\} \mid k \in \mathbb{N}\cap[0,r[, \omega_k \in]0,1], \\ u_k = \text{Im}(u_k), \mid u_k \mid = 1\} \qquad [4.8]$$

DEFINITION 4.9.– we define a series of increasing products:

$$\forall K \in \mathbb{N}\cap[0, 2^r[, \zeta(K, \Omega_r) := \eta(\beta(K), \Omega_r) \\ = \prod_{i=0}^{r-1} \exp(b_{r-1-i,K} 2\pi\omega_i u_i) \qquad [4.9]$$

DEFINITION 4.10.– we define a sample family, for r = 6,

$$\Omega_6{}^\dagger := \{\{\frac{1}{3},(\frac{\mathbb{i}+\mathbb{j}+\mathbb{k}}{\sqrt{3}})\}, \{\frac{1}{3},(\frac{\mathbb{i}+\mathbb{j}+\mathbb{k}}{\sqrt{3}})\}, \{\frac{1}{6},(\frac{\mathbb{i}+\mathbb{j}+\mathbb{k}}{\sqrt{3}})\}, \{\frac{1}{4},(\frac{\mathbb{i}}{\sqrt{1}})\}, \{\frac{1}{4},(\frac{\mathbb{j}}{\sqrt{1}})\}, \{\frac{1}{8},(\frac{\mathbb{i}}{\sqrt{1}})\}\} \qquad [4.10]$$

Figure 4.7 shows that the simplest closed lattice of triple-binary trees with $\Omega_6{}^\dagger$ is made of two modules (see Wolfrma) and has 144 connections and 96 nodes.

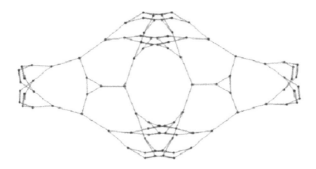

Figure 4.7. *The simplest closed lattice of triple-binary trees with $\Omega_6{}^t$ is made of two modules and has 144 connections and 96 nodes*

DEFINITION 4.11.– we define a unitary diagonal quaternion:

$$\mathbb{m}: = (\frac{\hat{\imath}+\hat{\jmath}+\hat{k}}{\sqrt{3}}) \tag{4.11}$$

DEFINITION 4.12.– we define a suite of quaternions:

$$\zeta(K,\Omega_6{}^t) := ((((\exp(b_{5,K}\frac{2\pi}{3}\mathbb{m})\exp(b_{4,K}\frac{2\pi}{3}\mathbb{m}))\exp(b_{3,K}\frac{2\pi}{6}\mathbb{m}))$$

$$\exp(b_{2,K}\frac{2\pi}{4}\hat{\imath}))\exp(b_{1,K}\frac{2\pi}{4}\hat{\jmath}))\exp(b_{0,K}\frac{2\pi}{8}\hat{\imath}) \tag{4.12}$$

DEFINITION 4.13.– a field $\mathbb{E}$ is defined at each node, having a quaternionic value computed using two functions, ω and ζ, and one set of constants Ω_r:

$$\forall K \in \mathbb{N}\cap[0,2^{r+1}[,\forall n_k \in \text{nodes}(\mathbb{B}_{r+1}),$$

$$\mathbb{E}(n_k): = \begin{cases} \text{if} k < 2^r: \mathbb{E}(n_{2k}) \\ \text{else: } \omega(\zeta(K-2^r,\Omega_r))\frac{\text{Im}(\zeta(K-2^r,\Omega_r))}{|\text{Im}(\zeta(K-2^r,\Omega_r))|} \end{cases} \tag{4.13}$$

DEFINITION 4.14.– we define a reordering function (see Figure 4.8):

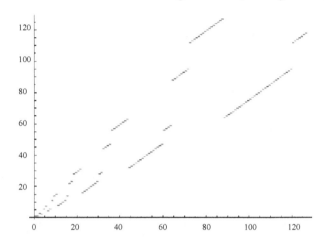

Figure 4.8. *Plot of the α (n) function (the values of α vs. n)*

$$\alpha(K) := ((K - 2^{\lfloor\frac{LogK}{Log2}\rfloor-1} + 5 * 2^{\lfloor\frac{LogK}{Log2}\rfloor-3})[6 * 2^{\lfloor\frac{LogK}{Log2}\rfloor-3}]$$
$$- 2^{\lfloor\frac{LogK}{Log2}\rfloor-2})[2^{\lfloor\frac{LogK}{Log2}\rfloor}] + 2^{\lfloor\frac{LogK}{Log2}\rfloor-1} \qquad [4.14]$$

DEFINITION 4.15.– we define the lattice of connected triple-binary graphs, modulo x_0:

$$\mathbb{L}(\mathbb{T}_{r+1}, x_0) :=$$
$$\bigcup(\mathbb{T}_r(x))\bigcup\bigcup_{n_k\in nodes(\mathbb{T}_{r+1}), k\geqslant 2^r, k<11*2^{r-3}}(n_k(\mathbb{T}_{r+1}(x)), n_{\alpha(k)}(\mathbb{T}_{r+1}(x + 2^{\frac{k[2]}{2}}\exp(\mathbb{E}(n_k(\mathbb{T}_{r+1}(x))))[x_0]))) \quad [4.15]$$

DEFINITION 4.16.– we define a lattice condition:

Lattice condition on $\mathbb{E}$ (therefore on Ω_r): $\forall n_k \in nodes(\mathbb{T}_{r+1})$,
$$k \geqslant 2^r, k < 11 * 2^{r-3}, \mathbb{R}^4(2^{\frac{2+k[2]}{2}}\exp(\mathbb{E}(n_k(\mathbb{T}_{r+1}(x))))) \in \mathbb{N}^4 \qquad [4.16]$$

THEOREM 4.1.– lattice condition equation [4.16] holds for $\Omega_6{}^\dagger$.

DEMONSTRATION.– The values taken by $\mathbb{E}$ on the 48 leaves are the 24 units of the Hurwitz integer, forming the binary tetrahedral group, and their products by (1+i). They are also coordinates of the 24-cell [BOO 10] and its dual scaled by square root of two (see Table 4.1). They form the integer lattice F4, of the roots of Lie algebra ≻4.

VALUES OF $\mathbb{E}(n_k)$ FIELD							
K	$\{\zeta(K,\Omega_6{}^\dagger)\}$	ω_k	$\{u_k\}$	K	$\{\zeta(K,\Omega_6{}^\dagger)\}$	ω_k	$\{u_k\}$
1	$\{1,0,0,0\}$	0	$\{0,0,0,0\}$	25	$\{-1,0,0,0\}$	$\frac{1}{2}$	$\{0,\frac{1}{2\sqrt{3}},\frac{1}{2\sqrt{3}},\frac{1}{2\sqrt{3}}\}$
2	$\{\frac{1}{\sqrt{2}},\frac{1}{\sqrt{2}},0,0\}$	$\frac{1}{8}$	$\{0,\frac{1}{8},0,0\}$	26	$\{-\frac{1}{\sqrt{2}},-\frac{1}{\sqrt{2}},0,0\}$	$\frac{3}{8}$	$\{0,-\frac{3}{8},0,0\}$
3	$\{0,0,1,0\}$	$\frac{1}{4}$	$\{0,0,\frac{1}{4},0\}$	27	$\{0,0,-1,0\}$	$\frac{1}{4}$	$\{0,0,-\frac{1}{4},0\}$
4	$\{0,0,\frac{1}{\sqrt{2}},-\frac{1}{\sqrt{2}}\}$	$\frac{1}{4}$	$\{0,0,\frac{1}{4\sqrt{2}},-\frac{1}{4\sqrt{2}}\}$	28	$\{0,0,-\frac{1}{\sqrt{2}},\frac{1}{\sqrt{2}}\}$	$\frac{1}{4}$	$\{0,0,-\frac{1}{4\sqrt{2}},\frac{1}{4\sqrt{2}}\}$
5	$\{0,1,0,0\}$	$\frac{1}{4}$	$\{0,\frac{1}{4},0,0\}$	29	$\{0,-1,0,0\}$	$\frac{1}{4}$	$\{0,-\frac{1}{4},0,0\}$
6	$\{-\frac{1}{\sqrt{2}},\frac{1}{\sqrt{2}},0,0\}$	$\frac{3}{8}$	$\{0,\frac{3}{8},0,0\}$	30	$\{\frac{1}{\sqrt{2}},-\frac{1}{\sqrt{2}},0,0\}$	$\frac{1}{8}$	$\{0,-\frac{1}{8},0,0\}$
7	$\{0,0,0,1\}$	$\frac{1}{4}$	$\{0,0,0,\frac{1}{4}\}$	31	$\{0,0,0,-1\}$	$\frac{1}{4}$	$\{0,0,0,-\frac{1}{4}\}$
8	$\{0,0,\frac{1}{\sqrt{2}},\frac{1}{\sqrt{2}}\}$	$\frac{1}{4}$	$\{0,0,\frac{1}{4\sqrt{2}},\frac{1}{4\sqrt{2}}\}$	32	$\{0,0,-\frac{1}{\sqrt{2}},-\frac{1}{\sqrt{2}}\}$	$\frac{1}{4}$	$\{0,0,-\frac{1}{4\sqrt{2}},-\frac{1}{4\sqrt{2}}\}$
9	$\{\frac{1}{2},\frac{1}{2},\frac{1}{2},\frac{1}{2}\}$	$\frac{1}{6}$	$\{0,\frac{1}{6\sqrt{3}},\frac{1}{6\sqrt{3}},\frac{1}{6\sqrt{3}}\}$	49	$\{-\frac{1}{2},-\frac{1}{2},-\frac{1}{2},-\frac{1}{2}\}$	$\frac{1}{3}$	$\{0,-\frac{1}{3\sqrt{3}},-\frac{1}{3\sqrt{3}},-\frac{1}{3\sqrt{3}}\}$
10	$\{0,\frac{1}{\sqrt{2}},\frac{1}{\sqrt{2}},0\}$	$\frac{1}{4}$	$\{0,\frac{1}{4\sqrt{2}},\frac{1}{4\sqrt{2}},0\}$	50	$\{0,-\frac{1}{\sqrt{2}},-\frac{1}{\sqrt{2}},0\}$	$\frac{1}{4}$	$\{0,-\frac{1}{4\sqrt{2}},-\frac{1}{4\sqrt{2}},0\}$
11	$\{-\frac{1}{2},-\frac{1}{2},\frac{1}{2},\frac{1}{2}\}$	$\frac{1}{3}$	$\{0,-\frac{1}{3\sqrt{3}},\frac{1}{3\sqrt{3}},\frac{1}{3\sqrt{3}}\}$	51	$\{\frac{1}{2},\frac{1}{2},-\frac{1}{2},-\frac{1}{2}\}$	$\frac{1}{6}$	$\{0,\frac{1}{6\sqrt{3}},-\frac{1}{6\sqrt{3}},-\frac{1}{6\sqrt{3}}\}$
12	$\{0,-\frac{1}{\sqrt{2}},\frac{1}{\sqrt{2}},0\}$	$\frac{1}{4}$	$\{0,-\frac{1}{4\sqrt{2}},\frac{1}{4\sqrt{2}},0\}$	52	$\{0,\frac{1}{\sqrt{2}},-\frac{1}{\sqrt{2}},0\}$	$\frac{1}{4}$	$\{0,\frac{1}{4\sqrt{2}},-\frac{1}{4\sqrt{2}},0\}$
13	$\{-\frac{1}{2},\frac{1}{2},\frac{1}{2},-\frac{1}{2}\}$	$\frac{1}{3}$	$\{0,\frac{1}{3\sqrt{3}},\frac{1}{3\sqrt{3}},-\frac{1}{3\sqrt{3}}\}$	53	$\{\frac{1}{2},-\frac{1}{2},-\frac{1}{2},\frac{1}{2}\}$	$\frac{1}{6}$	$\{0,-\frac{1}{6\sqrt{3}},-\frac{1}{6\sqrt{3}},\frac{1}{6\sqrt{3}}\}$
14	$\{-\frac{1}{\sqrt{2}},0,0,-\frac{1}{\sqrt{2}}\}$	$\frac{3}{8}$	$\{0,0,0,-\frac{3}{8}\}$	54	$\{\frac{1}{\sqrt{2}},0,0,\frac{1}{\sqrt{2}}\}$	$\frac{1}{8}$	$\{0,0,0,\frac{1}{8}\}$
15	$\{-\frac{1}{2},\frac{1}{2},-\frac{1}{2},\frac{1}{2}\}$	$\frac{1}{3}$	$\{0,\frac{1}{3\sqrt{3}},-\frac{1}{3\sqrt{3}},\frac{1}{3\sqrt{3}}\}$	55	$\{\frac{1}{2},-\frac{1}{2},\frac{1}{2},-\frac{1}{2}\}$	$\frac{1}{6}$	$\{0,-\frac{1}{6\sqrt{3}},\frac{1}{6\sqrt{3}},-\frac{1}{6\sqrt{3}}\}$
16	$\{-\frac{1}{\sqrt{2}},0,0,\frac{1}{\sqrt{2}}\}$	$\frac{3}{8}$	$\{0,0,0,\frac{3}{8}\}$	56	$\{\frac{1}{\sqrt{2}},0,0,-\frac{1}{\sqrt{2}}\}$	$\frac{1}{8}$	$\{0,0,0,-\frac{1}{8}\}$
17	$\{-\frac{1}{2},\frac{1}{2},\frac{1}{2},\frac{1}{2}\}$	$\frac{1}{3}$	$\{0,\frac{1}{3\sqrt{3}},\frac{1}{3\sqrt{3}},\frac{1}{3\sqrt{3}}\}$	57	$\{\frac{1}{2},-\frac{1}{2},-\frac{1}{2},-\frac{1}{2}\}$	$\frac{1}{6}$	$\{0,-\frac{1}{6\sqrt{3}},-\frac{1}{6\sqrt{3}},-\frac{1}{6\sqrt{3}}\}$
18	$\{-\frac{1}{\sqrt{2}},0,\frac{1}{\sqrt{2}},0\}$	$\frac{3}{8}$	$\{0,0,\frac{3}{8},0\}$	58	$\{\frac{1}{\sqrt{2}},0,-\frac{1}{\sqrt{2}},0\}$	$\frac{1}{8}$	$\{0,0,-\frac{1}{8},0\}$
19	$\{-\frac{1}{2},-\frac{1}{2},-\frac{1}{2},\frac{1}{2}\}$	$\frac{1}{3}$	$\{0,-\frac{1}{3\sqrt{3}},-\frac{1}{3\sqrt{3}},\frac{1}{3\sqrt{3}}\}$	59	$\{\frac{1}{2},\frac{1}{2},\frac{1}{2},-\frac{1}{2}\}$	$\frac{1}{6}$	$\{0,\frac{1}{6\sqrt{3}},\frac{1}{6\sqrt{3}},-\frac{1}{6\sqrt{3}}\}$
20	$\{0,-\frac{1}{\sqrt{2}},0,\frac{1}{\sqrt{2}}\}$	$\frac{1}{4}$	$\{0,-\frac{1}{4\sqrt{2}},0,\frac{1}{4\sqrt{2}}\}$	60	$\{0,\frac{1}{\sqrt{2}},0,-\frac{1}{\sqrt{2}}\}$	$\frac{1}{4}$	$\{0,\frac{1}{4\sqrt{2}},0,-\frac{1}{4\sqrt{2}}\}$
21	$\{-\frac{1}{2},-\frac{1}{2},\frac{1}{2},-\frac{1}{2}\}$	$\frac{1}{3}$	$\{0,-\frac{1}{3\sqrt{3}},\frac{1}{3\sqrt{3}},-\frac{1}{3\sqrt{3}}\}$	61	$\{\frac{1}{2},\frac{1}{2},-\frac{1}{2},\frac{1}{2}\}$	$\frac{1}{6}$	$\{0,\frac{1}{6\sqrt{3}},-\frac{1}{6\sqrt{3}},\frac{1}{6\sqrt{3}}\}$
22	$\{0,-\frac{1}{\sqrt{2}},0,-\frac{1}{\sqrt{2}}\}$	$\frac{1}{4}$	$\{0,-\frac{1}{4\sqrt{2}},0,-\frac{1}{4\sqrt{2}}\}$	62	$\{0,\frac{1}{\sqrt{2}},0,\frac{1}{\sqrt{2}}\}$	$\frac{1}{4}$	$\{0,\frac{1}{4\sqrt{2}},0,\frac{1}{4\sqrt{2}}\}$
23	$\{-\frac{1}{2},\frac{1}{2},-\frac{1}{2},-\frac{1}{2}\}$	$\frac{1}{3}$	$\{0,\frac{1}{3\sqrt{3}},-\frac{1}{3\sqrt{3}},-\frac{1}{3\sqrt{3}}\}$	63	$\{\frac{1}{2},-\frac{1}{2},\frac{1}{2},\frac{1}{2}\}$	$\frac{1}{6}$	$\{0,-\frac{1}{6\sqrt{3}},\frac{1}{6\sqrt{3}},\frac{1}{6\sqrt{3}}\}$
24	$\{-\frac{1}{\sqrt{2}},0,-\frac{1}{\sqrt{2}},0\}$	$\frac{3}{8}$	$\{0,0,-\frac{3}{8},0\}$	64	$\{\frac{1}{\sqrt{2}},0,\frac{1}{\sqrt{2}},0\}$	$\frac{1}{8}$	$\{0,0,\frac{1}{8},0\}$

Field values for the 48 leaves of the triple binary tree $\mathbb{T}_7 = \Xi(\mathbb{B}_7)$ based on $\Omega_6{}^\dagger$, pointing toward linked supernode, such that opposite leaves have opposite field.

Table 4.1. *Values of $\mathbb{E}(n_k)$ field*

A large family of graphiton models can be built by varying r, Ω_r and x_0, where equation [4.16] holds.

Figure 4.9 shows that the second simplest closed lattice of triple-binary trees with $\Omega_6{}^\dagger$ is made of 32 modules and has 2,304 connections [WOL 02].

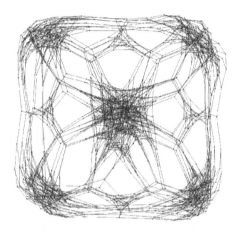

Figure 4.9. *The second simplest closed lattice of triple-binary trees with $\Omega_6{}^\dagger$ is made of 32 modules and has 2,304 connections*

4.4. Perturbation tolerance

DEFINITION 4.17.– we define a bit value at each node:

$$\forall K \in \mathbb{N} \cap [0, 2^r[, \forall n_k \in \text{nodes}(\mathbb{T}_{r+1}),\ \ \mathbb{I}(n_k) := k[2] \tag{4.17}$$

Topological bits in a triple-binary tree according to our method preserve the intrinsic geometry. The center of the super node is determined as the single loop of size 4. Associated with the adjacent single triangle, they form the central triangle of three nodes respectively informed of bits 0, 0 and 1, and each connected to a first node of binary tree associated respectively with 0, 1 and 1. Thus, each tree root is characterized by a binary code on two different bits, and then the subsequent nodes of the branches at each level are identified by 0 and 1. And the 24 leaves of three binary trees of rank 6 are characterized by a 5-bit identifier [FEM 94]. The lattice condition equation [4.16] expresses that the identifiers of two connected leaves issued from two connected triple trees have binary identifiers correlated by the function α that introduces redundant information and thus tolerance to disturbance.

4.5. Conclusion

Research in the field of artificial intelligence and attempts at architectures of neural networks were conducted confidentially with hyperdiamond and hypercube structures. Indeed, given that the nature of space-time is linked to four-dimensional geometry, spatial artificial intelligence, even in space-time, has good reason to rely on these geometric structures. We believe that graphiton models deserve to be studied in this area.

Pancake and hypercube networks, hypercube and in particular the cross hypercube [TSA 08] are the subject of active research to optimize reliability, transmission delay, rate and energy consumption in communication networks, particularly new networks with dynamic topology.

Our graphiton models can be considered as hybrid networks. They seem particularly interesting when the large-scale infrastructure is fixed; it is only on a small scale that nodes in the network connect or disconnect dynamically. This behavior corresponds to swapping bits by means of the T1 transformations (as shown in Figure 4.4) and local changes in the network structure. We believe that the imminent arrival of the Internet of Things and the increasing use of RFID tags, and integrated NFC antennae in mobile phones will result in the emergence of hybrid physical networks. Graphiton models could be useful in the simulation and management of these new hybrid networks. Our main result is that these models define an information structure which encodes both the program and data, the hardware would be managed by a local graph operator analyzing the surrounding environment and deciding whether or not to make a T1 topological transformation. This dynamic graph is a promising new application in artificial intelligence and the optimization of networks.

It was clarified that the proposed architecture is suitable for the production of four-dimensional neural networks and the recognition of 3D trajectories.

The application for RFID tags is dynamically linked to a smartphone equipped with detectors and a geolocation system, this enables the processing of the space-time dynamics of this information. RFID tags would be modeled by empty supernodes of our hyperdiamond network activated when detected. The features of this application will be discussed in the future coming work.

4.6. Bibliography

[ALE 26] ALEXANDER J.W., BRIGGS G.B., "On types of knotted curves", *Annals of Mathematics*, 2nd Series, vol. 28, nos. 1–4 (1926/27), pp. 562–586, ISSN 0003-486X, 1926.

[AMB 04] AMBJØRN J., JURKIEWICZ J., LOLL R., "Emergence of a 4D world from causal quantum gravity", *Physical Review Letters*, vol. 93, p. 131–301, 2004, available at: http://arxiv.org/abs/hep-th/0404156.

[ASC 08] ASCHHEIM R., "Bitmaps for a digital theory of everything", *Midwest NKS Conference, Indiana University*, USA, available at: http://www.cs.indiana.edu/~dgerman/2008midwestNKSconference/rasch.pdf, 31 October–02 pp. 1–30, November 2008.

[ASC 09a] ASCHHEIM R., "From NKS to E8 symmetry, a description of the universe", *JOUAL '09*, CNR, Pisa, Italy, 10 July 2009.

[ASC 09b] ASCHHEIM R., "Hyperdiamant", *From the Earth to the Stars Symposium*, Ars Mathematica, Metz, France, 16–22 November, 2009.

[ASC 10] ASCHHEIM R., Graphitation, digital relativity, NKS Summer School, Vermont University, USA, 9 July 2010.

[BOO 10] BOOLE STOTT A., "Geometrical deduction of semiregular from regular polytopes and space fillings", *Verhandelingen der Koninklijke Akademie van Wetenschappente Amsterdam*, vol. 11, no. 1, pp. 3–24, 2010.

[CAR 10] CARLIP S., "The small scale structure of spacetime", in ELLIS G., MURUGAN J., WELTMAN A. (eds), *Foundations of Space and Time*, Cambridge University Press, Cambridge, UK, available at: http://arxiv.org/abs/1009.1136, 2010.

[CHA 10] CHAMSEDDINE A., CONNES A., "Noncommutative geometry as a framework for unification of all fundamental interactions including gravity. Part I", *Fortschritte der Physik*, vol. 58, pp. 553–600, available at: http://arxiv.org/abs/1004.0464v1, 2010.

[CON 90] CONNES A., *Géométrie non commutative*, Inter-Editions, Paris, 1990.

[DE 19] DE OCKAM G., "Quaestiones et decisiones in quatuor libros Sententiarum cum centilogio theologico", Book II, available at: http://fondotesis.us.es/fondos/libros/256/553/quaestiones-et-decisiones-in-iv-libros-sententiarum-cum-centilogio-theologico/?desplegar=2708, 1319.

[DE 40] DE OCKAM G., "Dialogus", Book 3, Part 3, tract 2, Chapter 17, available at: http://www.britac.ac.uk/pubs/dialogus/t32d3c3.html, 1340.

[FEM 94] FEMMAM S., Study and design of lattice networks algorithms applied to adaptive prediction filters, Master Thesis, University of Annaba, Algeria, June 1994.

[KON 08] KONOPKA T., MARKOPOULOU F., SEVERINI S., "Quantum graphity: a model of emergent locality", *Physics Review*, vol. D77, p. 104029, available at: http://arxiv.org/abs/0801.0861, 2008.

[LUM 01] LUMINET J.P., *L'univers chiffonné*, Fayard, Paris, 2001.

[MAR 04] MARKOPOULOU F., SMOLIN L., "Quantum theory from quantum gravity", *Physical Review*, vol. D70, p. 124029, available at: http://arxiv.org/abs/gr-qc/0311059, 2004.

[PEN 05] PENROSE R., *The Road to Reality: A Complete Guide to the Laws of the Universe*, Knopf, New York, 2005.

[SAD 97] SADOC J.F., MOSSERI R., *Frustration géométrique*, Eyrolles, Paris, 1997.

[TSA 08] TSAI P.-Y., FU J.-S., CHEN G.-H., "Edge-fault-tolerant Hamiltonicity of pancake graphs under the conditional fault model", *Theoretical Computer Science*, vol. 409, pp. 450–460, 2008.

[WOL 02] WOLFRAM S., *A New Kind of Science*, Wolfram Media, Champaign, USA, 2002.

Beacon Cluster-Tree Construction for ZigBee/IEEE802.15.4 Networks

5.1. Introduction

In this chapter, we discuss sensor networks, an infrastructure composed of sensing, computing and communication elements that enable an administrator to observe and react to events and phenomena in a specified environment [ILY 05]. This kind of network is characterized by low energy consumption [DON 08], low rate and low cost. Applications based on Wireless Sensor Networks (WSN) are growing every day, and they cover diversified application fields such as ambient assisted living, building automation and factory automation.

The IEEE 802.15.4 standard defines the physical layer and the medium access control (MAC) sub-layer for low-rate wireless area networks (LR-WPAN) [IEE 06].

The IEEE 802.15.4 MAC sub-layer allows two modes for transmitting and receiving data: the beacon-enabled mode and the non-beacon-enabled mode.

The former is a synchronized mode; it can guarantee transmission determinism within Guaranteed Time Slots (GTSs). The latter does not give any traffic guarantee and does not need any synchronization between network devices.

Chapter written by Smain FEMMAM, Mohamed Ikbal BENAKILA and Laurent GEORGE.

Three topologies are available in the IEEE 802.15.4: Mesh, Star and Cluster-Tree. The beacon mode has been designed to work with a star topology. However, no mechanisms have been defined in the IEEE 802.15.4 standard to enable the beacon mode using a Mesh topology, and it is not clearly defined how to establish a beacon Cluster-Tree topology. One of the advantages of WSNs is that they allow the establishment of large networks with a considerable amount of nodes (up to 65,000 nodes for ZigBee standard) covering large areas [GUN 09, YOO 10]. In this context, beacon Cluster-Tree networks, such as network grants, are most suitable due to the GTS mechanism, a certain QoS for time-sensitive applications (see [CHA 10] and [JUN 10]).

However, constructing a beacon Cluster-Tree network is not simple, because the beacon mode is a SuperFrame-based one. For a given network node, the SuperFrame starts when the beacon frame is received from the parent node. Thus, in the presence of several beacon coordinators (in the case of large WSNs), beacon Cluster-Tree construction mechanisms should avoid the collision of beacon frames.

Several works have been conducted to enable the construction of large WSNs. We have proposed in Benakila et al. [BEN 10] the definition of a new device, called the beacon-aware device. This device allows the cohabitation of beacon and non-beacon networks. The beacon-aware devices permit the creation of a network comprising both beacon and non-beacon devices. The solution presented in Benakila et al. [BEN 10] guarantees the integrity of the beacon network traffic by introducing a channel access priority mechanism. In addition to the network size and topology, the QoS is an important parameter to take into consideration. Some sensor applications that require a bounded transport delay can use the Guaranteed Time Slot (GTS) mechanism defined by the IEEE 802.15.4 standard within a fully beacon network. In the work of Jurcik et al. [JUR 08] and Femmam and Benakila [FEM 16], the authors proposed a modeling methodology for Cluster-Tree networks in order to compute worst-case end-to-end delay and buffering and bandwidth requirements. This modeling method enables the network designers to create Cluster-Tree networks that fit with their application constraints.

The IEEE 802.15.4/ZigBee standard defines the Cluster-Tree topology as a special case of a peer-to-peer network. However, the realization of beacon Cluster-Tree networks is not defined in the standard. Some work has been done in order to model Cluster-Tree topologies [MHA 04, KOU 06, KOU 07] and failure recovery [GUP 03] and to allow the construction of beacon Cluster-Tree networks.

The RFC submitted to the Task Group 15.4b (see the work of T.G.15.4b [IEE 04]) proposes enhancements to the IEEE 802.15.4 standard. The construction of beacon Cluster-Tree topologies was one of the document topics. The authors of the work of T.G.15.4b classified beacon frame conflicts into two categories: direct conflict and indirect conflict. They proposed some approaches to solve each category of beacon conflict. We present the beacon conflict categories and the approaches proposed by the work of T.G.15.4b in section 5.4. Koubaa, Cunha and Alves proposed a SuperFrame scheduling algorithm to enhance the approach introduced in the work of T.G.15.4b. Indeed, the authors of the work of T.G.15.4b did not introduce any scheduling algorithm. The authors of Junghee reference proposed a method to maximize the lifetime of a Cluster-Tree ZigBee network under the constraint that every data flow should deterministically meet its end-to-end deadline for hard real-time sensing/control applications. The authors of Koubaa et al. [KOU 06, KOU 07] tackled the problem by introducing the constraints and the algorithm needed to provide a strong scheduling mechanism. The objective is to meet all end-to-end deadlines of a predefined set of time-bounded data flows while minimizing the energy consumption of the nodes [HAN 10]. The authors of Francomme et al. [FRA 07] proposed a mechanism to schedule beacon frame transmissions called the beacon-only period approach. In this approach, there is no need to schedule SuperFrames, that is, all the SuperFrames start at the same time. Only beacon frame transmissions are scheduled within a new SuperFrame period called the beacon-only period (for more details, see section 5.4 of this chapter). In addition, the authors of Francomme et al. [FRA 07] introduced a GTS collision avoidance mechanism, which guarantees certain traffic QoS. The authors of Radeke et al. [RAD 08] focused on achieving the node mobility management strategy.

In this chapter, all the network devices will transmit during the same SuperFrame, which means that the beacon order (BO) and the SuperFrame order (SO) parameters will be the same for the whole network. Nevertheless, using our approach, SuperFrames and beacon frame scheduling will be avoided and the construction of a Beacon Cluster-Tree network will be possible without introducing any constraints on the SuperFrame parameters.

The rest of the chapter is organized as follows: the next section presents a brief overview of the IEEE 802.15.4 MAC sub-layer. Section 5.3 introduces the beacon collision problem within a Cluster-Tree topology. In section 5.4, we present our approach to resolve the problem. A narrowband multipath fading model is presented in section 5.5. Section 5.6 contains a MATLAB/Simulink simulation model of multipath impacts on wireless reception. Test bed bench results are presented in section 5.7, and finally, we conclude this chapter.

5.2. IEEE 802.15.4 overview

In this section, we give an overview of the MAC sub-layer and the physical layer specification of IEEE 802.15.4.

5.2.1. *EEE 802.15.4 physical layer*

The main functions of the physical layer are spreading, de-spreading, modulation and demodulation of the signal. The IEEE 802.15.4 defines three frequency bands, in which LRWPANs could be deployed: 868 and 915 MHz and 2.4 GHz. This chapter focuses on the 2.4 GHz frequency band, because it has been used worldwide. Table 5.1 gives a summary of the 2.4 GHz physical layer properties.

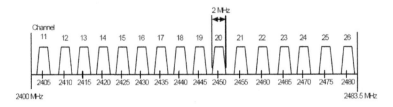

Figure 5.1. *IEEE 802.15.4 channel selection*

MHz	Frequency Band	Chip rate (kchip/s)	Modulation	Bit rate (kb/s)	Symb rate (ksym/s)
2450	2400–2483.5	2000	OQPSK	250	62.5

Table 5.1. *IEEE 802.15.4 physical layer parameters*

The 2.4 GHz frequency band contains 11 channels numbered from Ch 11 to Ch 26; Figure 5.1 shows the channel selection process. IEEE 802.15.4 uses the direct sequence spread spectrum (DSSS) as the spreading technique. DSSS is used to increase the frequency of the signal in order to increase its power and reduce the influence of noise. The 2.4 GHz band uses the offset quadrature phase shift keying (OQPSK) technique for chip modulation and DSSS. Each 4-bit symbol is mapped into a 32-chip PN (pseudo-random noise) sequence. The chip sequences representing each data symbol are modulated onto the carrier using OQPSK with half-sine pulse shaping. Even-indexed chips are modulated onto the I-phase (I) carrier and odd-indexed chips are modulated onto the quadrature-phase (Q) carrier. To form the offset between the I-phase and Q-phase chip modulation, the Q-phase chips shall be delayed by Tc with respect to the I-phase chips, where Tc is the inverse of the chip rate. Receiver and transmitter parameters are defined by the standard; some of these parameters are listed in the following table.

Transmit power	Receiver sensitivity	Receiver bandwidth
0 dbm	−85 dbm	2 MHz

Table 5.2. *IEEE 802.15.4 TRX parameters*

5.2.2. IEEE 802.15.4 MAC sub-layer overview

The IEEE 802.15.4 standard is a suitable protocol for low-rate wireless networks. A lot of effort has been made to make the standard low power and self-organized.

Two kinds of devices have been introduced in wireless medium access control (MAC) and physical layer (Phy) specifications for low-rate wireless personal area networks (WPANs): reduced function devices (RFD) and full function devices (FFD). An FFD implements all the functions defined by the standard. However, an RFD implements the basic functions (joining a network, leaving a network, transmitting, etc.) defined in the standard. Three topologies are allowed by the standard.

5.2.2.1. *Mesh topology*

In a mesh topology, there is one coordinator and a set of nodes associated to it. Each node is a router and permits other nodes to associate. A given node can communicate directly with other nodes if they are in its POS (Personal Operating Space, i.e. in-range transmission), or passing by other nodes (acting as routers in this case) to reach its target node (see Figure 5.2). Using this topology, no synchronization is needed between the devices. Furthermore, enabling the synchronization in such a topology can be problematic, as synchronization mechanisms for mesh topology are not defined in the standard.

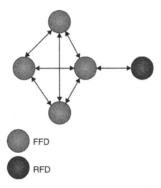

Figure 5.2. *Mesh topology*

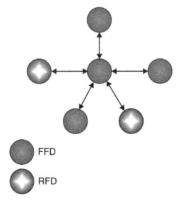

Figure 5.3. *Star topology*

5.2.2.2. *Star topology*

The coordinator is the main node in the network. All other nodes must be associated to it, and all communications between nodes must pass through it, even if the communication initiator node and the target node are in the POS of each other (see Figure 5.3). Performing synchronization (beacon mode) using this topology has been well defined by the IEEE 802.15.4 standard. A third topology, the Cluster-Tree topology, can be considered, too. This topology is very interesting for time-sensitive applications. Hereafter, the Cluster-Tree topology is introduced.

5.2.2.3. *Cluster-Tree topology*

A Cluster-Tree network is the one in which most devices are FFDs. An RFD connects to a cluster-tree network as a leaf device at the end of a branch (RFDs do not allow other devices to associate). An FFD device may act as a coordinator and provide synchronization services to other devices or other coordinators. The PAN coordinator forms the first cluster by choosing an unused PAN identifier and broadcasting beacon frames to neighboring devices.

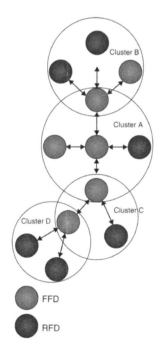

Figure 5.4. *Cluster-Tree topology*

A candidate device receiving a beacon frame may request to join the network at the PAN coordinator. If the PAN coordinator permits the device to join, it adds the new device as a child device to its neighbor list. Then, the newly joined device adds the PAN coordinator as its parent to its neighbor list and begins transmitting periodic beacons; other candidate devices may then join the network at that device. The simplest form of a cluster-tree network is a single-cluster network, but larger networks are possible by forming a mesh of multiple neighboring clusters. Once predetermined application or network requirements are met, the first PAN coordinator may instruct a device to become the PAN coordinator of a new cluster (Cluster head) adjacent to the first one. Other devices gradually connect and form a multi-cluster network structure (see Figure 5.4).

5.2.3. *Non-beacon-enabled network*

This mode assumes that every node can communicate directly with other nodes without any synchronization requirements. A node can transmit and go to sleep at any time following its own energy consumption policy. All transmissions are done after performing the unslotted CSMA/CA algorithm, to check if the channel is clear for transmission [BUR 09, BUR 10]. A non-beacon device transmits the beacon frame only as a response to a beacon request command. Devices operating in this mode do not need to synchronize with other devices.

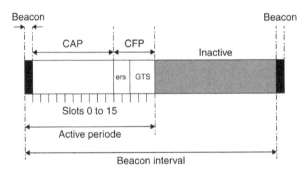

Figure 5.5. *SuperFrame structure*

5.2.4. Beacon-enabled network

In a beacon-enabled mode, the coordinator plays a crucial role. It defines periods of time in which transmissions can be made and intervals of time where all nodes associated to it must go to sleep. In this mode, the time is divided into a succession of SuperFrames. A SuperFrame is a time interval that contains an active period and an inactive period. The Beacon Interval (BI) parameter indicates the interval between two successive beacon frames. The length of the active period is indicated by the SD (SuperFrame Duration) parameter. The active period is divided into a fixed number of 16 time slots of equal sizes. All beacon network communications are carried out within this period. The active period is divided into a contention access period (CAP) and a contention-free period (CFP) [BUR 09, BUR 10].

The CAP is the period where all nodes compete for channel access using the slotted CSMA/CA algorithm.

The CFP gathers GTSs (Guaranteed Time Slots). A GTS is one or more slots of time reserved for a particular node. A GTS is unidirectional, that is, only for receptions or transmissions. The coordinator starts allocating GTSs from the last time slot to the first slot respecting a maximum size of the CFP. GTS transmissions do not need the use of the CSMA/CA algorithm for channel access, because the slots are reserved for one node. The structure of a SuperFrame is illustrated in Figure 5.5.

The beacon mode forces all the devices to synchronize with the coordinator. This is done by the reception and the process of the beacon frame. The most important parameters for synchronization are the BO, for computing the BI, the SuperFrame Order (SO), for computing the SuperFrame duration (SD) and the final CAP slot parameter. The final CAP slot indicates the end of the CAP. After this slot, only devices owning a GTS can transmit. Figure 5.6 presents the format of the beacon frame. The BO is a parameter used by the associated nodes for calculating the beacon interval, which is computed using the following formula:

$$BI = aBaseSuperFrameDuration \times 2^{BO} \qquad\qquad [5.1]$$

with *aBaseSuperFrameDuration* = 60 symbols.

The BO value should be between 0 and 14. A BO of 15 indicates that the device operates in the non-beacon mode. The SO is a parameter used for calculating the active period duration:

$$SD = aBaseSuperFrameDuration \times 2^{SO}$$ [5.2]

with $0 \leq SO \leq BO \leq 14$

The authors of Buratti et al. [BUR 09, BUR 10] provided a mathematical model that helps network designers in choosing the BO and SO parameters.

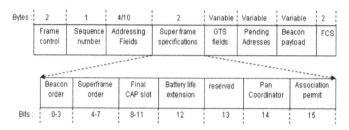

Figure 5.6. *Beacon frame fields*

5.3. Beacon frame collision problem in a Cluster-Tree topology

Beacon frames are transmitted without the CSMA/CA algorithm and hence the risk of beacon collisions still remains [KIM 08]. This problem has been addressed as a request for comment (RFC) by the Task Group 15.4b. Two types of beacon frame collision have been identified in T.G.15.4b.

5.3.1. *Direct beacon frame collision*

A direct beacon frame collision occurs when more than one coordinator is in the transmission range of one another, and the beacon transmission occurs in approximately the same time (see Figure 5.7(a)). Two main approaches have been proposed to solve the problem of direct beacon frame collision: the time division approach and the beacon-only period approach. These approaches were defined in T.G.15.4b and developed in Koubaa et al. [KOU 06, KOU 07] and Francomme et al. [FRA 07].

5.3.1.1. *The time division approach*

The time division approach consists of the following principles: a given coordinator transmits its beacon frame and spends its active period during the inactive period of its neighboring coordinators. The Task Group 15.4b did not propose any scheduling algorithm in order to increase the efficiency of the mechanism. This lack has been tackled in Koubaa *et al.* [KOU 06, KOU 07]. The authors of Koubaa *et al.* [KOU 06, KOU 07] proposed a SuperFrame scheduling algorithm in order to maximize the number of clusters in the network. This approach suffers from several problems:

– To enable parent–child communications, a coordinator is activated during its active period and the active period of its parent coordinator.

– The increasing density of devices in the network makes the problem more complicated, and the scheduling algorithm may return an "un-schedulable set" response, which means that the Cluster-Tree topology cannot be used.

– To make the SuperFrame scheduling algorithm more efficient, the authors of Koubaa *et al.* [KOU 06, KOU 07] have made restrictive constraints on the SO and BO parameters, which could perturb the execution of some applications in the network.

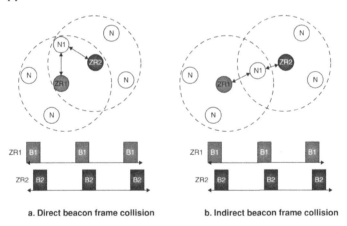

a. Direct beacon frame collision   b. Indirect beacon frame collision

Figure 5.7. *Direct and indirect beacon frame collisions*

5.3.1.2. The beacon-only period approach

In the beacon-only period approach, the SuperFrame structure is modified. A time period, called "Beacon-Only period", is added at the beginning of the SuperFrame. The beacon-only period is divided into time slots called "Contention-Free Time Slots" (CFTS). Each coordinator transmits its beacon frame within its CFTS (see Figure 5.8). Thus, beacon frame collisions will be avoided. This approach was first presented by the Task Group 15.4b and then developed by Koubaa *et al.* [KOU 06, KOU 07] and Francomme *et al.* [FRA 07].

However, dimensioning the beacon-only period is complicated, because the duration of the period must be evaluated dynamically depending on the association and leaving actions of beacon coordinators. In addition, the beacon-only approach does not avoid indirect beacon frame collisions, or it does, but, by including the global CFTS scheduling algorithm.

5.3.2. Indirect beacon frame collision

An indirect beacon frame collision occurs when a given node is in the transmission range of two or more coordinators. The coordinators send their beacon frames at approximately the same time; that is, at a given time, the node receives more than one beacon frame, which results in the collision. This situation is a typical hidden nodes transmission situation (see Figure 5.7(b)). Indirect beacon frame collisions can be solved by the following two approaches.

5.3.2.1. The reactive approach

The network is started normally and the coordinators do not do much to prevent the beacon frames from colliding. Once a collision occurs, the node (the node in the POS of more than one coordinator, see Figure 5.7) will start orphan scans to try to re-synchronize with its coordinator.

However, if after a number of orphan scans, the node is still unable to receive the beacon frame correctly, it initiates a beacon conflict command; beacon coordinators receiving this command will adjust their beacon transmission time in order to solve the problem.

This approach is simple, but the recovery from a beacon conflict can take a long time.

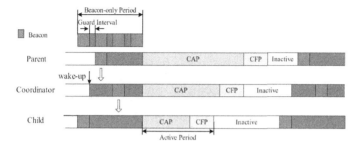

Figure 5.8. *Beacon-only period approach*

5.3.2.2. *The proactive approach*

In the proactive approach, coordinators try to avoid beacon frame conflicts before starting their beacon frame transmissions. A beacon coordinator listens to the channel and collects its neighbor's beacon frame transmission time.

However, if a beacon frame collision is reported, the network is able to solve the problem using the reactive approach. It is important to note that this approach is more complicated than the first one.

5.4. Proposed new approach

In this section, we propose a new approach that allows forming Cluster-Tree networks without regard to the density of the network. Our approach is not based on SuperFrames or even Beacon frame transmission scheduling.

In our approach, we do not introduce changes on the way an 802.15.4 Cluster-Tree network is constructed. We mainly focus on node synchronization and frame transmission in a beacon Cluster-Tree without introducing scheduling algorithms. This approach is based on the following:

– The same BO and SO values are used for all the nodes of the network (WPANs).

– All the nodes are synchronized thanks to beacon frame transmissions.

– All the nodes transmit during the same SuperFrame.

To enable collision-free beacon transmissions, we adopt a novel approach, described in this section.

The beacon frame is divided into two parts:

– The common part: It is the part that does not change from a beacon coordinator to another. It contains the SO and BO parameters.

– The specific part: It is the part specific to a beacon coordinator, that is, changes from a given beacon coordinator to another.

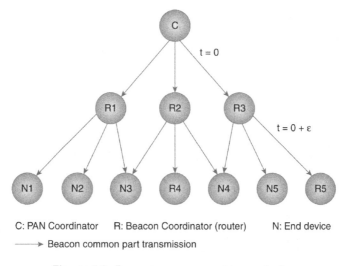

C: PAN Coordinator   R: Beacon Coordinator (router)   N: End device
——→ Beacon common part transmission

Figure 5.9. *Beacon common part transmission*

Each part is a separate frame. The transmission mechanism is described in Figures 5.9 and 5.10.

Node synchronization is achieved by the transmission of the common part. It gives the start signal to sensor nodes to begin the beacon mode SuperFrame. The common part contains only synchronization information (BO and SO). The specific part is transmitted during the CAP period, and it contains the traditional beacon information (GTSs organization, Pending addresses, etc.).

The data included in the specific part are specific to each beacon coordinator. The specific part is transmitted by each beacon coordinator during the CAP using the slotted CSMA/CA algorithm to access the channel as specified in the IEEE 802.15.4 standard. Therefore, as all the nodes share the same SuperFrame and the specific part is transmitted by performing a channel access algorithm, collisions between neighbor beacon coordinators should be reduced and treated as normal data transmission. However, this approach affects the orphan detection mechanism. A given node should consider itself as an orphan node if it does not receive the specific part for *aMaxLostBeacons* times.

The PAN Coordinator broadcasts the beacon's common part. When the beacon's common part is received, a beacon coordinator begins its SuperFrame and forwards the same frame (i.e. the beacon's common part) to its neighbor nodes. Using this mechanism, beacon coordinators at the same level of the Cluster-Tree transmit the common part at the same time. This should not cause a reception problem if a node receives more than one frame at the same time.

Indeed, all the beacon routers are broadcasting the same frame, or the same bit configuration, which means that all the beacon routers are transmitting the same RF signal, which can be assimilated to a multipath reception.

Multipath propagation occurs when RF signals take different paths from a source to a destination. A part of the signal goes to the destination while another part bounces off an obstruction and then arrives at the destination. As a result, part of the signal encounters delay and travels a longer path to the destination. Multipath can be defined as the combination of the original signal and the duplicate wave fronts that result from the waves' reflection off obstacles between the transmitter and the receiver CISCO [CIS 08].

Multipath propagation occurs even with only one transmitter and one receiver. Nowadays, receivers are able to retrieve the information from an RF signal perturbed by multipath signals, because interference, and not multipath fading, is the primary cause of unpredictable performance [RAM 09].

Thus, a node receiving the beacon common part from more than one beacon coordinator (i.e. more than one RF signal) should be able to retrieve the common part information, as the node is able to deal with multipath Radio Frequency (RF) signals.

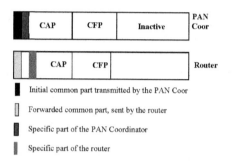

Figure 5.10. *Beacon specific part transmission*

5.5. Narrowband multipath fading model

If a single pulse is transmitted over a multipath channel, then the received signal will appear as a pulse train, with each pulse in the train corresponding to the line-of-sight (LOS) component or a distinct multipath component associated with a distinct scatter. The time delay spread of a multipath channel can result in significant distortion of the received signal.

We say that two multipath components with delay and τ_2 are resolvable if their delay difference significantly exceeds the inverse signal bandwidth: $|\tau_1 - \tau_2| \gg 1/B_u$. Multipath components that do not satisfy this resolvability criterion cannot be separated out at the receiver; thus, these components are non-resolvable. These non-resolvable components are combined into a single multipath component, with amplitude and phase corresponding to the sum of the different components [GOL 05]. In this chapter, we assume that the resolvability criterion is satisfied. Indeed, the transmission range of an IEEE 802.15.4 is short; thus, transmission delays between two signals with different origin nodes should respect the criterion.

Another characteristic of the multipath channel is the time-varying nature. Time variation arises because either the transmitter or the receiver is moving and hence the location of reflectors in the transmitting path will change over time [GOL 05].

Let the transmitted signal be:

$$s(t) = \text{Re}\{u(t)e^{j2\pi f_c t}\} = \text{Re}\{u(t)\}\cos(2\pi f_c t) - \text{Im}\{u(t)\}\sin(2\pi f_c t) \qquad [5.3]$$

where $u(t)$ is the equivalent low-pass signal for $s(t)$ with bandwidth B_u and f_c is the carrier frequency. Neglecting noise, the received signal is the sum of the line-of-sight path and all resolvable multipath components [GOL 05]:

$$r(t) = \text{Re}\left\{\sum_{n=0}^{N(t)}\alpha_n(t)u(t-\tau_n(t))e^{j(2\pi f_c(t-\tau_n(t))+\varphi_{D_n}(t))}\right\} \qquad [5.4]$$

where:

n = 0 is the LOS component.

$N(t)$ is the number of the resolvable multipath components.

$\tau_n(t)$ is the delay of the nth multipath component. It depends on the path length $r_n(t)$ and could be written as: $\tau_n(t) = r_n(t) \div c$

$\alpha_n(t)$ is the amplitude. $\varphi_{D_n}(t)$ is the Doppler phase shift.

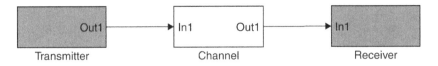

Figure 5.11. *Simulink simulation model*

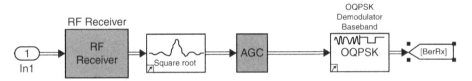

Figure 5.12. *Receiver architecture*

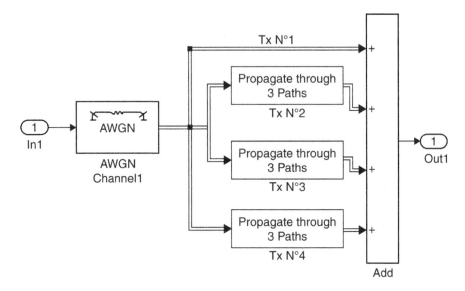

Figure 5.13. *Channel architecture*

We assume that all the nodes in the network are static; thus, $\varphi_{D_n}(t) = 0$.

For simplification, let (as IEEE 802.15.4 is a narrowband channel) the delay spread T_m satisfy $T_m \ll B^{-1}$. Thus, $\tau_i \leq T_m$ for all i, so $u(t - \tau_i(t)) \approx u(t)$.

Equation [5.4] could be written as:

$$r(t) = \text{Re}\left\{ u(t)e^{j2\pi f_c t} (\sum_{n=0}^{N(t)} \alpha_n(t)e^{-j\varphi_n(t)}) \right\} \qquad [5.5]$$

The complex scale factor within large parentheses is independent of the transmitted signal and the equivalent low-pass signal $u(t)$. If we consider $s(t)$

to be an unmodulated carrier with random phase offset φ_0, then s(t) could be written as:

$$s(t) = \mathrm{Re}\left\{ e^{j(2\pi f_c t + \varphi_0)} \right\}$$ [5.6]

The received signal $r(t)$ will be:

$$r(t) = \mathrm{Re}\left\{ \left(\sum_{n=0}^{N(t)} \alpha_n(t) e^{-j\varphi_n(t)} \right) e^{j2\pi f_c t} \right\}$$ [5.7]

$$r(t) = r_I(t)\cos 2\pi f_c t - r_Q(t)\sin 2\pi f_c t$$ [5.8]

where:

$$r_I(t) = \sum_{n=0}^{N(t)} \alpha_n(t)\cos\varphi_n(t)$$ [5.9]

$$r_Q(t) = -\sum_{n=0}^{N(t)} \alpha_n(t)\sin\varphi_n(t)$$ [5.10]

And the phase $\varphi_n(t)$ is:

$$\varphi_n(t) = 2\pi f_c \tau_n(t) - \varphi_0$$ [5.11]

Now, considering P transmitters transmitting the same RF signal, the received signal $r(t)$ is:

$$r(t) = r_I(t)\cos 2\pi f_c t - r_Q(t)\sin 2\pi f_c t$$ [5.12]

where:

$$r_I(t) = \sum_{l=0}^{P} \sum_{n=0}^{N(t)} \alpha_{ln}(t)\cos\varphi_{ln}$$ [5.13]

$$r_Q(t) = \sum_{l=0}^{P} \sum_{n=0}^{N(t)} \alpha_{ln}(t)\sin\varphi_{ln}(t)$$ [5.14]

$\alpha_{\mathrm{l}n}(t)$ is the amplitude of the nth multipath component transmitted by the lth transmitter.

$\varphi_n(t)$ is the phase of the nth multipath component transmitted by the lth transmitter.

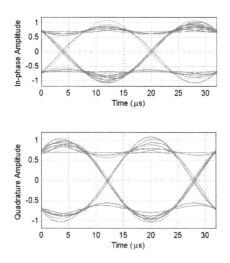

Figure 5.14. *LOS transmission scatter plot*

5.6. Model multipath simulation

In this section, a MATLAB/Simulink model is developed to test the mechanism and measure the impact of a multipath reception on the BER value.

The model is shown in Figure 5.11; it consists of a transmitter, a channel and a receiver. The transmitter is sending an OQPSK modulated signal over a channel constituted of an AWGN channel and a collection of transmitting paths. Each path is emulating an independent transmitter, and each transmitted signal is propagated through three paths with variable delays, thus simulating multipath phenomena (see Figure 5.13). All transmitted signals and multipaths are combined at the receiver (see Figure 5.12), which filters the signal in order to retrieve the information.

Figures 5.14 and 5.15 show, respectively, the scatter plot and the eye diagram obtained from this simulation in the case of the presence of only an LOS signal. In order to measure the impact of receiving a signal from several transmitters on the BER value, we assume one LOS transmitter and we increase the number of non-LOS transmitters. For each number of non-LOS transmitters, we measure the BER value at several noise levels. Each transmitter line contains three delayed versions of the signal.

Figure 5.16 shows the BER results for one transmitter (LOS transmitter), two transmitters (one LOS transmitter and a non-LOS transmitter), three transmitters (one LOS transmitter and two non-LOS transmitters) and four transmitters (one LOS transmitter and three non-LOS transmitters).

From Figure 5.16, we can clearly see that, in the presence of an LOS component, the number of non-LOS transmitters does not significantly affect the BER value.

As a second step, we are interested in measuring the BER value in the presence of several LOS transmitters. In this simulation, we measure the BER value for each number of present transmitters and at several noise levels. Figure 5.17 shows the BER values obtained from this simulation, from which we can conclude that the presence of more than one LOS transmitter affects the BER value significantly, which means that to guarantee an acceptable BER value, an LOS transmitter should be present.

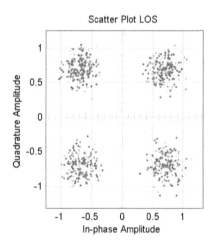

Figure 5.15. *LOS eye diagram*

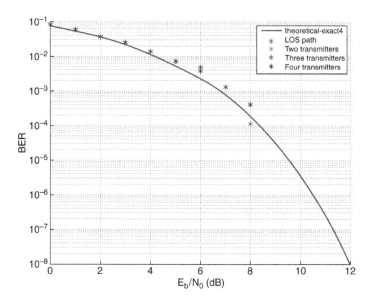

Figure 5.16. *BER in the presence of several transmitters but only one LOS transmitter*

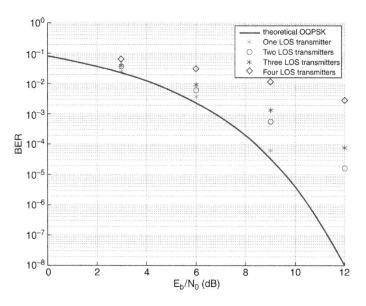

Figure 5.17. *BER in the presence of several LOS transmitters*

5.7. Test bed bench

The core of the presented approach is the exploitation of multipath phenomena to avoid beacon transmission scheduling. Assuming that RF receivers are able to deal with multipath signals to retrieve the information, they can retrieve the information when receiving an RF signal transmitted by different nodes at the same time, because each transmitted signal will be treated as a multipath signal by the receiver.

To stress this point, we conceived a real experimental setup using the TelosB motes. Crossbow's TelosB mote (Figure 5.18) is an open-source platform designed to enable cutting-edge experimentation for the research community.

These notes offer:

– an IEEE 802.15.4/ZigBee compliant RF transceiver;

– a globally compatible ISM band;

– a 250 kbps data rate;

– low power consumption.

Figure 5.18. *Crossbow's TelosB mote*

The principle is to send the same frame by several nodes to one receiver. For visual consideration, we chose a MicroChip sniffer as a second receiver.

An example of the network architecture with three routers is presented in Figure 5.19. The receiver is in the transmission range of the PAN Coordinator and the three routers, that is, the sniffer's software show the frames transmitted by the coordinator and the routers. When a router receives the frame from the PAN Coordinator, it retransmits it immediately, that is, all routers transmit the same frame at the same time.

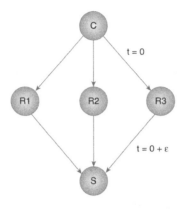

C: PAN Coordinator   R: Beacon Coordinator (router)   S: Sniffer

———➤ Beacon common part transmission

Figure 5.19. *A test network architecture with three routers*

As shown in Figure 5.19, the sniffer receives two frames: the first is the frame transmitted by the PAN Coordinator and the second is the frame transmitted by the routers.

The sniffer receives only one frame from the routers even if there is more than one transmission. Consequently, the receiver considers all transmissions as one transmission (see Figure 5.20).

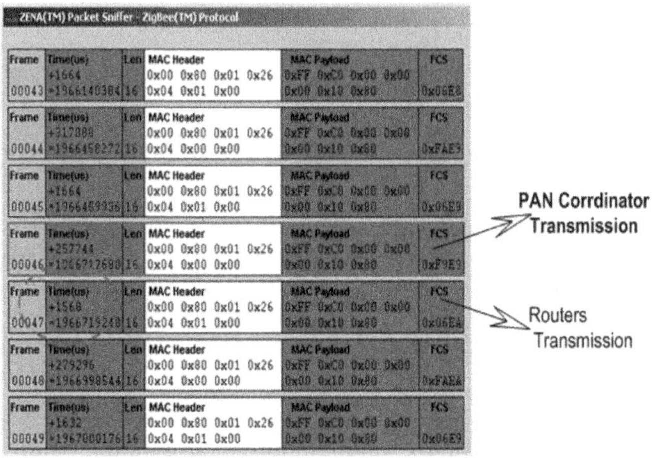

Figure 5.20. *Packets received from the sniffer*

The receiver's capability to process multipaths avoids the introduction of beacon or SuperFrames scheduling mechanisms.

In order to measure the impact of the distance and the number of transmitting nodes on the beacon common part frame receiving mechanism, a test was conducted. It consists of a receiver node and a variable number of nodes at variable distances in an indoor (office) environment and an outdoor environment. For each measure, transmitter nodes transmit 500 beacon common frames.

These results are plotted in Figures 5.21 and 5.22 for indoor and outdoor environments, respectively. From these figures, we can clearly see that the packet loss is greater in the indoor environment than in the outdoor environment. This is a consequence of the number of multipath components arriving at the receiver node. Indeed, generally, scattering objects are less in outdoor environments than in indoor environments. In addition, the packet loss increases with the number of transmitters present.

The second test is deployed inside an office area; we fix the receiver's position and we vary the number of beacon's common part frame transmitters. Figure 5.23 shows a part of the floor map and node deployment (one receiver and 10 transmitters). Each beacon transmitter transmits 500 packets and we measure the number of packets correctly received by the receiver. The position of transmitter nodes is chosen randomly, and, for each number of transmitter nodes, several measures are made with different transmitter node positions.

Figure 5.24 shows the number of received beacon frames versus the number of transmitters present in the receiver's range. We can see that, in a real case, where beacon coordinators are deployed in an indoor environment, the receiver node is well tolerant to simultaneous transmissions, which makes this mechanism very interesting for constructing beacon Cluster-Tree topologies.

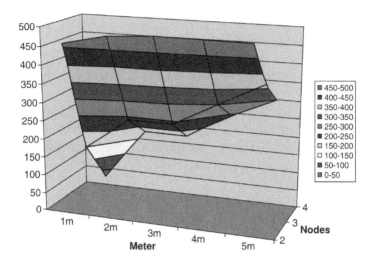

Figure 5.21. *Received common part packets versus distance and number of transmitters in an indoor environment. For a color version of the figure, see www.iste.co.uk/femmam/wireless.zip*

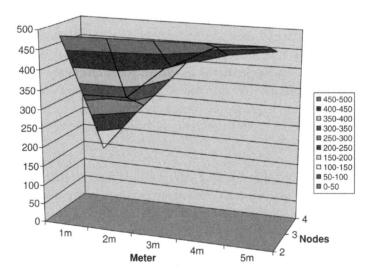

Figure 5.22. *Received common part packets versus distance and number of transmitters in an outdoor environment. For a color version of the figure, see www.iste.co.uk/femmam/wireless.zip*

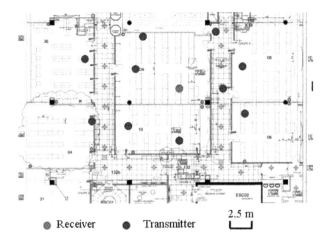

Figure 5.23. *Node deployment on the floor*

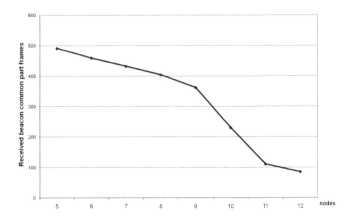

Figure 5.24. *BER versus the number of transmitter nodes in the area of the receiver*

5.8. Conclusion

In this chapter, we presented a new approach for the construction of ZigBee/IEEE 802.15.4 Cluster-Tree networks. This approach addresses the problems of beacon frame and SuperFrame scheduling and allows the construction of a Cluster-Tree topology, without introducing constraints on SuperFrame structure.

We proposed a collision-free beacon transmission approach that exploits the capabilities of the nodes to extract the information from a signal perturbed by simultaneous transmissions of several beacon coordinators. These transmitted RF signals are seen as reflected RF signals, because all the beacon coordinators are broadcasting the same signal. We show from experiments and simulations that a receiver is able to retrieve the information from a combination of several transmitted signals.

From experimental results, we can see that the proposed mechanism could be applied to beacon Cluster-Tree networks with a reduced number (until three) of beacon coordinators in a given geographical zone. More beacon coordinators could be tolerated in a given geographical zone if less QoS is needed on the network.

Future works will deal with adapting the presented approach to enable the construction of Beacon mesh networks. For time-sensitive applications, a GTS collision avoidance mechanism must be introduced to grant GTS traffic.

5.9. Bibliography

[BEN 10] BENAKILA M.I., FEMMAM S., GEORGE L., "Designing a ZigBee network for signal perception," *European Journal of Scientific Research*, vol. 40, no. 2, pp. 264–278, 2010.

[BUR 09] BURATTI C., VERDONE R., "Performance analysis of IEEE 802.15.4 non beacon-enabled mode", *IEEE Transactions on Vehicular Technology*, vol. 58, no. 7, pp. 3480–3493, Septemper 2009.

[BUR 10] BURATTI C., "Performance Analysis of IEEE 802.15.4 Beacon-Enabled Mode", *IEEE Transactions on Vehicular Technology*, vol. 59, no. 4, pp. 2031–2045, May 2010.

[CHA 10] CHALHOUB G., MISSON M., "Cluster-tree based energy efficient protocol for wireless sensor networks", *Proceeding ICNSC*, pp. 664–669, April 2010.

[CIS 08] CISCO, "Multipath and diversity", available at: http://www.cisco.com/application/pdf/paws/27147/multipath.pdf, 2008.

[DON 08] DONDI D., BERTACCHINI A., BRUNELLI D. *et al.*, "Modeling and optimization of a solar energy harvester system for self-powered wireless sensor networks," *IEEE Transactions on Industrial Electronics,* vol. 55, no. 7, pp. 2759–2766, July 2008.

[FEM 16] FEMMAM S., BENAKILA I.M., "A new topology time division beacon construction approach for IEEE802.15.4/ZigBee Cluster-tree Wireless Sensor Networks", *IEEE DataCom 2016, the 2nd IEEE International Conference on Big Data Intelligence and Computing*, Auckland, New Zealand, 8–12 August 2016.

[FRA 07] FRANCOMME J., MERCIER G., VAL T., "Beacon synchronization for gts collision avoidance in an ieee 802.15.4 meshed network," *Proceeding of IFAC*, November 2007.

[GOL 05] GOLDSMITH A., *Wireless Communications*, Cambridge University Press, 2005.

[GUN 09] GUNGOR V.C., HANCKE G.P., "Industrial wireless sensor networks: challenges, design principles, and technical approach," *IEEE Transactions on Industrial Electronics*, vol. 56, no. 10, pp. 4258–4265, October 2009.

[GUP 03] GUPTA G., Younis M., "Fault-tolerant clustering of wireless sensor networks", *Proceedings of WCNC*, pp. 1579–1584, March 2003.

[HAN 10] HANZALEK Z., JURCIK P., "Energy efficient scheduling for Cluster-Tree wireless sensor networks with time-bounded data flows: application to IEEE 802.15.4/ZigBee", *IEEE Transactions on Industrial Informatics*, vol. 6 issue 3, pp. 438–450, May. 2010.

[IEE 04] IEEE, T.G.15.4b, available at: http://grouper.ieee.org/groups/802/15/pub/tg4b.html, 2004.

[IEE 06] IEEE, Part 15.4: Wireless medium access control (MAC) and physical layer (Phy) specifications for low-rate wireless personal area networks (WPANs), IEEE Standard, 2006.

[ILY 05] ILYAS M., HAHGOUB I., *Handbook of Sensor Networks: Compact Wireless and Wired Sensing Systems*, CRC Press, 2005.

[JUN 09] JUNGHEE H., "Global Optimization of ZigBee parameters for end-to-end deadline guarantee of real-time data", *IEEE Sensors Journal*, vol. 9, no. 5, pp. 512–514, May 2009.

[JUN 10] JUNGHEE H., SUHAN C., TAEJOON P., "Maximizing lifetime of cluster-tree ZigBee networks under end-to-end deadline constraints", *IEEE Communications Letters*, vol. 14, no. 3, pp. 214–216, March 2010.

[JUR 08] JURCIK P., SEVERINO R., KOUBAA A. *et al.*, "Real-time communications over cluster-tree sensor networks with mobile sink behaviour," in *Proceedings of RTCSA*, pp. 401–411, August 2008.

[KIM 08] KIM J.-W., KIM J., EOM D.-S., "Multi-dimensional channel management scheme to avoid beacon collision in LR-WPAN", IEEE *Transactions on Consumer Electronics*, vol. 54 issue 2, pp. 396–404, May 2008.

[KOU 06] KOUBAA A., ALVES M., TOVAR E., "Modeling and worst-case dimensioning of cluster-tree wireless sensor networks", *Proceedings of RTSS*, pp. 412–421, December 2006.

[KOU 07] KOUBAA A., CUNHA A., ALVES M., "A time division beacon scheduling mechanism for IEEE 802.15.4/Zigbee Cluster-Tree Wireless Sensor Networks," *Proceedings of ECRTS*, pp.125–135, July 2007.

[MHA 04] MHATRE V., ROSENBERG C., "Design guidelines for wireless sensor networks: communication, clustering and aggregation", *Elsevier Ad Hoc Networks*, vol. 2, no. 1, pp. 45–63, January 2004.

[RAD 08] RADEKE R., MARANDIN D., CLAUDIOS F.J. *et al.*, "On reconfiguration in case of node mobility in clustered wireless sensor networks", *IEEE Wireless Communications*, vol. 15, no. 6, pp. 47–53, December 2008.

[RAM 09] RAMAN B., CHEBROLU K., GOKHALE D. *et al.*, "On the feasibility of the link abstraction in wireless mesh networks", *IEEE/ACM Transactions on Networking,* vol. 17, no. 2, pp. 528–541, April 2009.

[YOO 10] YOO S., CHONG P.K., KIM D. *et al.*, "Guaranteeing real-time services for industrial wireless sensor networks with IEEE 802.15.4," *IEEE Transactions on Industrial Electronics*, vol. 57, no. 11, pp. 3868–3876, November 2010.

6

One-by-One Embedding of the Twisted Hypercube into Pancake Graph

6.1. Introduction

Let G and H be two simple undirected graphs. An embedding of the graph G into the graph H is an injective mapping f from the vertices of G to the vertices of H. The study of graph embedding arises naturally in a number of computational problems: portability of algorithms across various parallel architectures and layout of circuits in VLSI. Akers and Krishnamurthy [AKE 89] proposed the pancake as an alternative to the hypercube and its variations for interconnecting processors in parallel computers.

This network has desirable properties: small diameter and fixed degree, $(n-1)$ regular, high connectivity, vertex symmetry, Hamiltonian, fault tolerance, extensibility and embeddability of other topologies cited in [HUN 03, HEY 97, ROW 93, ROW 98]. The embedding capabilities are important in evaluating an interconnection network. The embedding of the guest graph G into the host graph H is a mapping from each vertex of G to one vertex of H and mapping each edge of G to one path of H. Graph embedding is useful because an algorithm designed for H can be applied to G directly, and recent algorithms are cited in the references of this book [SEN 03, MEN 92, FAN 02, QIU 92, FAN 00, HSI 99, LIN 10, LIN 08]. Compared to the twisted hypercube, the pancake graph offers good

---

Chapter written by Smain FEMMAM and Faouzi M. ZERARKA.

and simple simulations of other interconnection networks, and the proofs are cited in the works of [MIL 94, SEN 97, HUN 02]. The basis of the mathematical theory used in this chapter can be found in [FEM 17].

This chapter is organized as follows. First, we introduce some definitions and notations, including the definition and properties of the twisted hypercube and the pancake network. In section 6.3, we present an algorithm of many-to-one embedding twisted hypercube into pancake. In section 6.4, we show that a dilation of many-to-one embedding of n-dimensional twisted embedding into pancake of dimension n is equal to 5. Finally, we conclude this chapter.

6.2. Preliminaries theory analysis

6.2.1. *Definition 6.1 construction*

The n-dimensional hypercube Q_n and the twisted hypercube $TQ_n = (V, U)$ have the same set of vertices V. We represent the address of each vertex in Q_n (TQ_n) as a binary string of length n. In such a way, we do not distinguish between vertices and their binary address. In Q_n, two vertices are adjacent if and only if their binary labels differ only in one bit position. For the TQ_n multiply twisted hypercube, adjacency requirements are a little more involved.

DEFINITION 6.1. Two binary strings $x = x_1x_0$ and $y = y_1y_0$ of length two are said to be pair-related if and only if $(x, y) \in \{(00, 00), (10, 10), (01, 11), (11, 01)\}$.

The n-dimensional multiply twisted hypercube TQ_n is recursively defined as follows: TQ_1 is the complete graph based on two vertices labeled 0 and 1. TQ_n consists of two sub-cubes $0TQ_{n-1}$ and $1TQ_{n-1}$, and the most significant bit of the labels of the vertices in $0TQ_{n-1}(1TQ_{n-1})$ is 0(1). U is the set of vertices $u = u_{n-1}u_{n-2}............u_1u_0 \in 0TQ_{n-1}$ with $u_{n-1} = 0$ and $v = v_{n-1}v_{n-2}............v_1v_0 \in 1TQ_{n-1}$ with $v_{n-1} = 1$ are joined by an edge in TQ_n if and only if:

$u_{n-2} = v_{n-2}$ if n is even   [6.1]

$(u_{2i+1}u_{2i}, u_{2i+1}u_{2i})$ are pair-related

Examples of twisted hypercube for $n = 1$, 2 and 3 are given in Figure 6.1.

The n-dimensional twisted hypercube TQ_n as an alternative to the hypercube has the same number of vertices V and degrees as the n-dimensional hypercube. The twisted hypercube is one of the variations of the hypercube, which is derived with some twisted edges. Because of these twisted edges, the diameter of TQ_n is only half that of the hypercube. Nice properties include relatively small degrees, embedding capabilities, scalability, robustness and the fault tolerance of Hamiltonicity of TQ_n as cited by [HUA 02, CHA 99, XU 10, ABR 91, HWA 00, HSI 98]. The twisted hypercube graph is not vertex-transitive for $n \geq 5$, as stated by [KUL 95].

6.2.2. Definition 6.2 construction

Cayley graphs were originally proposed as a generic theoretical model for analyzing symmetric interconnection networks. The most notable feature of the Cayley graph is its universality. The Cayley graph represents a class of high-performance interconnection network with a small degree and diameter, good connectivity and simple routing algorithms. The pancake is one of the Cayley graphs.

Let $I = (1,2,3,\ldots,n)$, $p = (p_1,p_2,\ldots\ldots p_n)$, $p_i \in I$ and $p_i \neq p_j$ for $i \neq j$, where p is the permutation of I.

A pancake graph $G_n = (P_n, E_n)$ of n dimensions is defined as follows:

$$P_n = \{(p_1, p_2...p_n) | \ p_i \in I, \ p_i \neq p_j \text{ for } i \neq j\} \qquad [6.2]$$

and:

$$E_n = \{((p_1, p_2,\ldots\ldots p_{j-1}, p_j, p_{j+1},\ldots\ldots\ldots, p_n), (p_j, p_{j-1},\ldots\ldots\ldots, p_2, p_1,$$

$$p_{j+1},\ldots\ldots\ldots, p_n)) | (p_1,p_2,\ldots\ldots p_n) \in P_n \text{ for } 2 \leq j \leq n. \qquad [6.3]$$

In other words, the set of P_n of all permutations constitutes the nodes of the vertices of G_n.

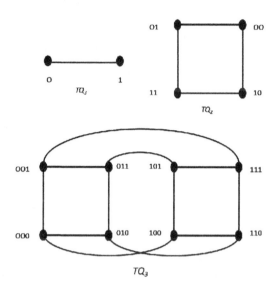

Figure 6.1. *Twisted hypercube for n = 1, 2 and 3*

Two nodes u and v are joined by an (undirected) edge if and only if the permutation corresponding to the node v can be obtained from u by flipping the object in positions 1 through j. Because for each permutation we can flip any number of objects between the first and j^{th} positions, $2 \leq j \leq n$, G_n is an $(n-1)$ regular graph, $|P_n| = n!$, $|E_n| = (n-1)n!/2$. Examples of pancake graphs for $n = 2$, 3 and 4 are given in Figures 6.2(a) and 6.2(b).

The pancake graphs proposed by Akers and Krishnamurthy [AKE 89] are an important family of interconnection networks. Some interesting properties of the pancake graphs are shown in Bouabdallah *et al.* [BOU 98] One of their main properties is that they are symmetric, as well as being built using Cayley groups with simple routing algorithms. Pancake graphs have many other attractive features, such as hierarchic, maximal fault tolerance and Hamiltonian, which are cited in [AKE 89, KAN 95, QIU 91], as well as small diameter, as shown in [MOR 96].

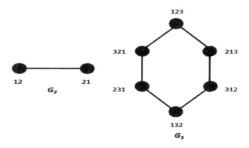

Figure 6.2(a). *Example of n-pancake graphs n = 2, 3*

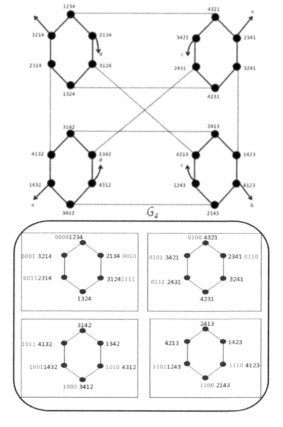

Figure 6.2(b). *Example of n-pancake graphs n = 4*

The graph G_n is made of n copies of G_{n-1}, namely $G_n[n, k]$ for $1 \leq k \leq n$. Considering each $G_n[n, k]$ as a super node, it follows that $G_n[n, s]$, $G_n[n, t]$ are connected by a collection of edges of the form $((t, p_2, p_3,...,p_{n-1},s)$, $(s, p_{n-1},...,p_2, t))$; thus, there are $(n-2)!$ edges connecting $G_n[n, s]$ and $G_n[n, t]$, as shown in [KAN 95] G_n is a complete graph on the super nodes connected by the super edges, as shown in Figure 6.3.

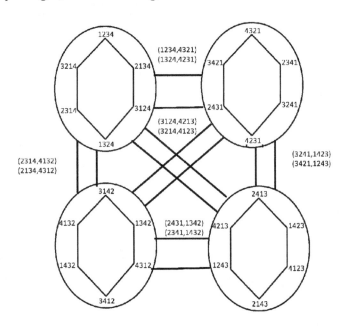

Figure 6.3. *Recursive structure of G_4*

6.2.3. *Definition 6.3 construction*

Let G and H be two simple undirected graphs. An embedding of the graph G into the graph H is an injective mapping f from the vertices of G to the vertices of H.

The dilation of the embedding is the maximum distance between $f(x)$ and $f(y)$ taken over all edges (x, y) of G.

6.2.4. *Notations*

A twisted hypercube of n dimensions is denoted by $TQ_n = (V, U)$, with V set of vertices and U set of edges.

A pancake of n dimensions is denoted by $G_n = (P_n, E_n)$, with P_n set of vertices and E_n set of edges.

$A \in V$ such that $A = a_1a_2a_3........a_{n-3}a_{n-2}a_{n-1}a_n = Prefa_{n-2}a_{n-1}a_n$, where $Pref = a_1a_2a_3........a_{n-3}$. $U1 \subset U$ as $u \in E1$, such that $u = (A, B)$ with A and B $\in V$.

$A \in V$, $A = a_1a_2a_3.....a_{n-4}.a_{n-3}a_{n-2}a_{n-1}a_n = Prefa_{n-4}a_{n-3}a_{n-2}a_{n-1}a_n$, where $Pref = a_1a_2a_3........a_{n-5}$.

$U2 \subset U$ as $u \in E2$, $u = (A, B)$ such that A and $B \in V$. $P_n' \subset P_n$ is a subset of P_n as $X \in P_n'$, where $X = x_1x_2x_3s_1s_2.....s_{n-l}$, such that $Suffix = s_1s_2........s_{n-l}$ and $l = (n-2)/2$, for $n > 3$.

$E_n' \subset E_n$ is a subset of paths, where all paths (X, Y) begin with X and end with Y, with $(X, Y) \in P_n'$.

$P_n'' \subset P_n$ is a subset of P_n, such that $X \in P_n''$, and $X = x_1x_2x_3x_4s_1s_2.....s_{n-l}$, such that the number of super node G_4 is equal to $l = (n-2)/2$, for $n > 4$, as $Suffix = s_1s_2.....s_{n-l}$.

$E_n'' \subset E_n$ is a subset of paths, where all paths (X, Y) begin with X and end with Y, with $(X, Y) \in P_n''$.

*Suffix*1(X) is a function which extracts the $n-3$ characters from a string X starting with the character of the lowest weight.

*Suffix*2(X) is a function which extracts the $n-4$ characters from a string X starting with the character of the lowest weight.

6.3. Embedding n-dimensional twisted hypercube graph into *n*-dimensional twisted pancake graph

In this section, we present a new function, the many-to-one embedding of an *n*-dimensional twisted hypercube graph denoted by TQ_n into an *n*-dimensional pancake graph denoted by G_n.

The main steps of the embedding function are as follows:

1) Find the first node of the twisted hypercube and the first node of the pancake. Example 000 of TQ_3 and 123 of G_3.

2) Embedding vertex of the twisted hypercube of *n* dimensions into the pancake of n dimensions using the Embed_node(node) algorithm.

3) Embedding edges of the twisted hypercube of *n* dimensions into the path of the pancake of *n* dimensions using the Embed_edge(nodedep, nodearr) algorithm.

6.3.1. *Embed_node(node) algorithm*

Embed_node(node) algorithm is done in the following way:

Case where *n*=3. Embedding the twisted hypercube of three dimensions into the pancake of three dimensions, as depicted in Figures 6.4 and 6.5.

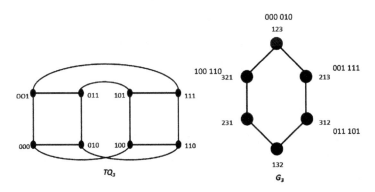

Figure 6.4. *Twisted hypercube and pancake of three dimensions*

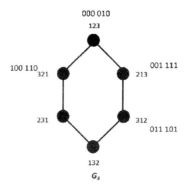

Figure 6.5. *Embedding graph of TQ$_3$ into G$_3$*

Generally, Embed_node(A) algorithm applies all actions specified in Table 6.1.

Nodes of TQ_n prefixed by 00	Nodes of $1^{st} G_n[n,n]$	Nodes of TQ_n prefixed by 10	Nodes of $2^{nd} G_n[n,1]$
00Pref000	$x_1x_2x_3Suf1$	*10Pref000*	$x_3x_2x_1Suf2$
00Pref001	$x_2x_1x_3Suf1$	*10Pref001*	$x_2x_3x_1Suf2$
00Pref010	$x_1x_2x_3Suf1$	*10Pref010*	$x_3x_2x_1Suf2$
00Pref011	$x_3x_1x_2Suf1$	*10Pref011*	$x_1x_3x_2Suf2$
00Pref100	$x_3x_2x_1Suf1$	*10Pref100*	$x_1x_2x_3Suf2$
00Pref101	$x_3x_1x_2Suf1$	*10Pref101*	$x_1x_3x_2Suf2$
00Pref110	$x_3x_2x_1Suf1$	*10Pref110*	$x_1x_2x_3Suf2$
00Pref111	$x_2x_1x_3Suf1$	*10Pref111*	$x_2x_3x_1Suf2$
Nodes of TQ_n prefixed by 01	**Nodes of $3^{rd} G_n[n, 3]$**	**Nodes of TQ_n prefixed by 11**	**Nodes of $4^{th} G_n[n, 2]$**
01Pref000	$x_3x_1x_2Suf3$	**11Pref000**	$x_2x_1x_3Suf4$
01Pref001	$x_3x_1x_2Suf3$	**11Pref001**	$x_1x_2x_3Suf4$
01Pref010	$x_3x_1x_2Suf3$	**11Pref010**	$x_2x_1x_3Suf4$
01Pref011	$x_1x_3x_2Suf3$	**11Pref011**	$x_2x_1x_3Suf4$
01Pref100	$x_1x_3x_2Suf3$	**11Pref100**	$x_1x_2x_3Suf4$
01Pref101	$x_1x_3x_2Suf3$	**11Pref101**	$x_2x_1x_3Suf4$
01Pref110	$x_1x_3x_2Suf3$	**11Pref110**	$x_1x_2x_3Suf4$
01Pref111	$x_3x_1x_2Suf3$	**11Pref111**	$x_1x_2x_3Suf4$

Table 6.1. *Embed_node(A) algorithm for*
A = A1PREF$A_{N-2}A_{N-1}A_N$, where A1 = 00, 01, 10, 11

The variable *Sufi* with ($i = 1...4$) is *Suffix*1(X), where $X \in P_i$, such that $G_n(n, k) = (P_i, E)$, where ($k = n, 1...3$).

Case where $n = 4$: The embedding nodes of TQ_4 in G_4 are produced as follows: TQ_4 is made recursively by two copies of TQ_3: one copy is prefixed by $0(0TQ_3)$ and the other one is prefixed by $1(1\ TQ_3)$. The G_4 is made recursively by four copies of G_3 named $G_4[4, k]$ for $k = 1...4$. We used in this case two copies of $G_4[4, k]$, for $k = 1, 4$. The first is $G_4[4, 4]$, used to embed all nodes of TQ_4 prefixed by $0(0\ TQ_3)$, and the second component $G_4[4, 1]$ is used to embed all nodes prefixed by $1(1\ TQ_3)$. The embedding is done by using the basic function of embedding of TQ_3 into G_3, as depicted in Figure 6.6. The embedding is done by using the rules specified in Table 6.2.

$0TQ_3$	$G_4[4, 4]$	$1TQ_3$	$G_4[4, 1]$
0000	$x_1x_2x_3x_4$	1000	$x_4x_3x_2x_1$
0001	$x_2x_1x_3x_4$	1001	$x_3x_4x_2x_1$
0010	$x_1x_2x_3x_4$	1010	$x_4x_3x_2x_1$
0011	$x_3x_1x_2x_4$	1011	$x_2x_4x_3x_1$
0100	$x_3x_2x_1x_4$	1100	$x_2x_3x_4x_1$
0101	$x_3x_1x_2x_4$	1101	$x_2x_4x_3x_1$
0110	$x_3x_2x_1x_4$	1110	$x_2x_3x_4x_1$
0111	$x_2x_1x_3x_4$	1111	$x_3x_4x_2x_1$

Table 6.2. *Embedding all nodes of TQ_4 into G_4*

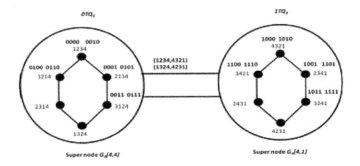

Figure 6.6. *Embedding graph of TQ_5 into G_5*

The case where $n = 5$: The embedded nodes of TQ_5 are produced as follows: TQ_5 is made recursively by prefixing the two copies of TQ_4: one with $0(0\ TQ_4)$ and the other $1(1\ TQ_4)$; in other words, $00TQ_3$, $01TQ_3$, $10TQ_3$, $11TQ_3$. The G_4 is made recursively by four copies of G_3 named $G_4[4, k]$, where $k = 1...4$. The first component is $G_4[4, 4]$ used for embedding nodes of TQ_5 prefixed by $00TQ_3$, the second is $G_4[4, 1]$ used for embedding all nodes $01TQ_3$, the third component is $G_4[4, 3]$ and the last component is $G_4[4, 2]$ used for embedding nodes of $11TQ_3$, as shown in Figure 6.7. The embedding is done using the rules specified in Table 6.3.

$00TQ_3$	$G_4[4, 4]$	$10TQ_3$	$G_4[4, 1]$	$01TQ_3$	$G_4[4, 3]$	$11TQ_3$	$G_4[4, 2]$
0000	$x_1x_2x_3x_4$	10000	$x_4x_3x_2x_1$	01000	$x_3x_4x_1x_2$	11000	$x_2x_1x_4x_3$
00001	$x_2x_1x_3x_4$	10001	$x_3x_4x_2x_1$	01001	$x_4x_3x_1x_2$	11001	$x_1x_2x_4x_3$
00010	$x_1x_2x_3x_4$	10010	$x_4x_3x_2x_1$	01010	$x_3x_4x_1x_2$	11010	$x_2x_1x_4x_3$
00011	$x_3x_1x_2x_4$	10011	$x_2x_4x_3x_1$	01011	$x_1x_3x_2x_4$	11011	$x_4x_2x_1x_3$
00100	$x_3x_2x_1x_4$	10100	$x_2x_3x_4x_1$	01100	$x_1x_4x_3x_2$	11100	$x_4x_1x_2x_3$
00101	$x_3x_1x_2x_4$	10101	$x_2x_4x_3x_1$	01101	$x_1x_3x_4x_2$	11101	$x_4x_2x_1x_3$
00110	$x_3x_2x_1x_4$	10110	$x_2x_3x_4x_1$	01110	$x_1x_4x_1x_2$	11110	$x_4x_1x_2x_3$
00111	$x_2x_1x_3x_4$	10111	$x_3x_4x_2x_1$	01111	$x_4x_3x_1x_2$	11111	$x_1x_2x_4x_3$

Table 6.3. *Embedding all nodes of TQ_5 into G_5*

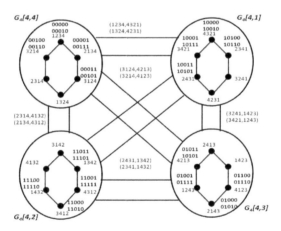

Figure 6.7. *Embedding graph of TQ_5 into G_5*

The case for $n > 5$: The twisted hypercube of n dimensions is produced by the composition of two copies of the twisted hypercube of $(n-1)$

dimensions. The first is prefixed by $0(0TQ_{n-1})$, and the second is prefixed by $1(1TQ_{n-1})$. The pancake of $n-1$ dimensions is made by I copies of $G_{n-1}[n-1, k]$, for $k=1, i$.

In other words, I super nodes containing 2^l components G_4, with $i = 2$ if n is even, $i = 4$ if n is odd and $l = (n-1)/2$.

There are two stated situations. The first one is, when n is even, we use two components: the super node $G_{n-1}[n-1, n-1]$ and the super node $G_{n-1}[n-1, 1]$, the first for embedding all nodes of $0TQ_{n-1}$ and the second one for embedding all nodes of $1TQ_{n-1}$. The second situation is when n is odd or $n = 2m+1(m \in \mathbb{N})$, the TQ_N nodes are $0TQ_{2m}$, $1TQ_{2m}$. For $N = 2m$, the TQ_N nodes are $0TQ_N$, $1TQ_N$, that is to say $00TQ_{N-1}$, $01TQ_{N-1}$ and $10TQ_{N-1}$, $11TQ_{N-1}$. In other words, we use four super nodes $G_{N-1}[N-1, N-1]$, $G_{N-1}[N-1, 1]$, $G_{N-1}[N-1, 2]$, $G_{N-1}[N-1, 3]$. The first node for embedding all nodes of $00TQ_{N-1}$, the second one for embedding all nodes of $10TQ_{N-1}$, the third one for embedding all nodes of $11TQ_{N-1}$ and the last node for embedding all nodes of $01TQ_{N-1}$ by using the rules specified in Table 6.4.

$APref00TQ_3$	$G_{n-1}[n-1, n-1]$	$APref10TQ_3$	$G_{n-1}[n-1, 1]$
$APref000000$	$x_1x_2x_3x_4suf1$	$APref100000$	$x_4x_3x_2x_1Suf2$
$APref000001$	$x_2x_1x_3x_4suf1$	$APref100001$	$x_3x_4x_2x_1Suf2$
$APref000010$	$x_1x_2x_3x_4Suf1$	$APref100010$	$x_4x_3x_2x_1Suf2$
$APref000011$	$x_3x_1x_2x_4suf1$	$APref100011$	$x_2x_4x_3x_1Suf2$
$APref000100$	$x_3x_2x_1x_4suf1$	$APref100100$	$x_2x_3x_4x_1Suf2$
$APref000101$	$x_3x_1x_2x_4suf1$	$APref100101$	$x_2x_4x_3x_1Suf2$
$APref000110$	$x_3x_2x_1x_4suf1$	$APref100110$	$x_2x_3x_4x_1Suf2$
$APref000111$	$x_2x_1x_3x_4suf1$	$APref100111$	$x_3x_4x_2x_1Suf2$
$APref01TQ_3$	$G_{n-1}[n-1, 3]$	$APref11TQ_3$	$G_{n-1}[n-1, 2]$
$APref010000$	$x_3x_4x_1x_2Suf3$	$APref110000$	$x_2x_1x_4x_3Suf4$
$APref010001$	$x_4x_3x_1x_2Suf3$	$APref110001$	$x_1x_2x_4x_3Suf4$
$APref010010$	$x_3x_4x_1x_2Suf3$	$APref110010$	$x_2x_1x_4x_3Suf4$
$APref010011$	$x_1x_3x_2x_4Suf3$	$APref110011$	$x_4x_2x_1x_3Suf4$
$APref010100$	$x_1x_4x_3x_2Suf3$	$APref110100$	$x_4x_1x_2x_3Suf4$
$APref010101$	$x_1x_3x_4x_2Suf3$	$APref110101$	$x_4x_2x_1x_3Suf4$
$APref010110$	$x_1x_4x_1x_2Suf3$	$APref110110$	$x_4x_1x_2x_3Suf4$
$APref010111$	$x_4x_3x_1x_2Suf3$	$APref110111$	$x_1x_2x_4x_3Suf4$

Table 6.4. Embedding all nodes of TQ_N IN G_5 for n>5

6.3.2. *Embed_edge(nodedep, nodearr) algorithm*

The Embed_edge(nodedep, nodearr) algorithm is given as follows:

```
Begin;
        S1:= Suffix1(nodedep);
    S2: = Suffix1(nodearr);
        S3:= Suffix2(nodedep);
    S4:= Suffix2(nodearr);
        If S1 = S2 then
                    Embed1_edge(nodedep, nodearr)
        Else
                If S3 = S4 then
                        Embed2_edge(nodedep, nodearr)

        Else
                        Embed3_edge(nodedep, nodarr)
                    Endif;
        Endif;
    End;
```

6.3.3. *Embed1_edge(nodedep, nodearr) algorithm*

The Embed1_edge(nodedep, nodearr) algorithm is used when the paths are in the same G_3 of a super node. This procedure applies the different cases outlined in Table 6.5 for A= 00 or 10, and the symmetric paths are shown in Table 6.6 for A= 01 or 11. It is important to note that the function *Suffix* is *Suffix2(X)*.

Twisted hypercube edge	Pancake path ($S1=Suffix$)	Dilation
APref000-APref001	$x_1x_2x_3x_4Suffix\text{-}x_2x_1x_3x_4Suffix$	1
APref000-APref010	$x_1x_2x_3x_4Suffix\text{-}x_1x_2x_3x_4Suffix$	1
APref000-APref100	$x_1x_2x_3x_4Suffix\text{-}x_3x_2x_1x_4Suffix$	1
APref001-APref011	$x_2x_1x_3x_4Suffix\text{-}x_3x_1x_2x_4Suffix$	1
APref001-APref111	$x_2x_1x_3x_4Suffix\text{-}x_2x_1x_3x_4Suffix$	1
APref010-APref011	$x_1x_2x_3x_4Suffix\text{-}x_2x_1x_3x_4Suffix\text{-}x_3x_1x_2x_4Suffix$	2
APref010-APref110	$x_1x_2x_3x_4Suffix\text{-}x_3x_2x_1x_4Suffix$	1
APref011-APref101	$x_3x_1x_2x_4Suffix\text{-}x_3x_1x_2x_4Suffix$	1
APref100-APref101	$x_3x_2x_1x_4Suffix\text{-}x_1x_2x_3x_4Suffix\text{-}x_2x_1x_3x_4Suffix\text{-}$ $x_3x_1x_2x_4Suffix$	3
APref101-APref111	$x_3x_1x_2x_4Suffix\text{-}x_2x_1x_3x_4Suffix$	1
APref110-APref111	$x_3x_2x_1x_4Suffix\text{-}x_1x_2x_3x_4Suffix\text{-}x_2x_1x_3x_4Suffix$	2

Table 6.5. *Embedding edges with label format 00Pref*
$X_1X_2X_3\text{–}00PREF\ Y_1Y_2Y_3$ *of twisted hypercube into pancake*

Twisted Hypercube edge	Pancake path ($S1=x_4Suffix$)	Dilation
APref000-APref001	$x_1x_3x_2x_4Suffix\text{-}x_3x_1x_2x_4Suffix$	1
APref000-APref010	$x_1x_3x_2x_4Suffix\text{-}x_1x_3x_2x_4Suffix$	1
APref000-APref100	$x_1x_3x_2x_4Suffix\text{-}x_2x_3x_1x_4Suffix$	1
APref001-APref011	$x_3x_1x_2x_4Suffix\text{-}x_2x_1x_3x_4Suffix$	1
APref001-APref111	$x_1x_3x_2x_4Suffix\text{-}x_1x_3x_2x_4Suffix$	1
APref010-APref011	$x_1x_3x_2x_4Suffix\text{-}x_3x_1x_2x_4Suffix\text{-}x_2x_1x_3x_4Suffix$	2
APref010-APref110	$x_1x_3x_2x_4Suffix\text{-}x_2x_3x_1x_4Suffix$	1
APref011-APref101	$x_2x_1x_3x_4Suffix\text{-}x_2x_1x_3x_4Suffix$	1
APref100-APref101	$x_2x_3x_1x_4Suffix\text{-}x_1x_3x_2x_4Suffix\text{-}x_3x_1x_2x_4Suffix\text{-}$ $x_2x_1x_3x_4Suffix$	3
APref101-APref111	$x_2x_1x_3x_4Suffix\text{-}x_1x_3x_2x_4Suffix$	1
APref110-APref111	$x_2x_3x_1x_4Suffix\text{-}x_1x_3x_2x_4Suffix\text{-}x_1x_3x_2x_4Suffix$	2

Table 6.6. *Embedding edges with label format 10Pref*
$X_1X_2X_3\text{–}10PREF\ Y_1Y_2Y_3$ *of twisted hypercube into pancake*

6.3.4. *Embed2_edge(nodedep, nodearr) algorithm*

This procedure is used when the paths are in the same G_4 of a super node. The Embed2_edge(nodedep, nodearr) algorithm realizes the embedding of the edge of twisted hypercube into the pancake, if the suffix of nodedep and the suffix of nodearr differ exactly in the fourth position.

Four cases arise in this situation. In the first case, the edge of the twisted hypercube is $Pref00a_{n-2}a_{n-1}a_n\text{-}Pref01a_{n-2}a_{n-1}a_n$, in the second it is $Pref00a_{n-2}a_{n-1}a_n\text{-}Pref10a_{n-2}a_{n-1}a_n$, in the third case it is $Pref01a_{n-2}a_{n-1}a_n\text{-}Pref11a_{n-2}a_{n-1}a_n$ and finally in the last case it is $Pref10a_{n-2}a_{n-1}a_n\text{-}Pref11a_{n-2}a_{n-1}a_n$.

The Embed2_edge(nodedep, nodearr) algorithm applies the actions outlined in Table 6.7.

Case	Twisted hypercube edge	Pancake path	Dilation
1	$Pref00000\text{-}Pref01000$ $Pref00010\text{-}Pref01010$	$x_1x_2x_3x_4Suffix\text{-}x_3x_2x_1x_4Suffix\text{-}$ $x_2x_1x_4x_3Suffix\text{-}x_2x_1x_4x_3Suffix$	3
1	$Pref00001\text{-}Pref01011$ $Pref00101\text{-}Pref01111$	$x_1x_2x_3x_4Suffix\text{-}x_2x_1x_3x_4Suffix\text{-}$ $x_4x_3x_2x_1Suffix$	2
1	$Pref00100\text{-}Pref01100$ $Pref00110\text{-}Pref01110$	$x_1x_2x_3x_4Suffix\text{-}x_4x_3x_2x_1Suffix$	1
2	$Pref00011\text{-}Pref01001$ $Pref00111\text{-}Pref01101$	$x_1x_2x_3x_4Suffix\text{-}x_4x_3x_2x_1Suffix\text{-}$ $x_2x_3x_4x_1Suffix$	2
2	$Pref00000\text{-}Pref10000$ $Pref00010\text{-}Pref10010$	$x_1x_2x_3x_4Suffix\text{-}x_4x_3x_2x_1Suffix$	1
2	$Pref00001\text{-}Pref10011$ $Pref00101\text{-}Pref10111$	$x_1x_2x_3x_4Suffix\text{-}x_4x_3x_2x_1Suffix\text{-}$ $x_2x_3x_4x_1Suffix\text{-}x_1x_4x_3x_2Suffix$	3
2	$Pref00100\text{-}Pref10100$ $Pref00110\text{-}Pref10110$	$x_1x_2x_3x_4Suffix\text{-}x_4x_3x_2x_1Suffix\text{-}$ $x_3x_4x_2x_1Suffix\text{-}x_1x_2x_4x_3Suffix$	3
2	$Pref00011\text{-}Pref10001$ $Pref00111\text{-}Pref10101$	$x_1x_2x_3x_4Suffix\text{-}x_3x_2x_1x_4Suffix\text{-}$ $x_4x_1x_2x_3Suffix\text{-}x_1x_4x_2x_3Suffix$	3

	Pref01000-Pref11000 *Pref01010-Pref11010*	$x_1x_2x_3x_4Suffix-x_4x_3x_2x_1Suffix$	1
	Pref01001-Pref11011 *Pref01101-Pref11111*	$x_1x_2x_3x_4Suffix-x_4x_3x_2x_1Suffix-$ $x_2x_3x_4x_1Suffix-x_1x_4x_3x_2Suffix$	3
3	*Pref01100-Pref11100* *Pref01110-Pref11110*	$x_1x_2x_3x_4Suffix-x_4x_3x_2x_1Suffix-$ $x_3x_4x_2x_1Suffix-x_1x_2x_4x_3Suffix$	3
	Pref01011-Pref11001 *Pref01111-Pref11101*	$x_1x_2x_3x_4Suffix-x_3x_2x_1x_4Suffix-$ $x_4x_1x_2x_3Suffix-x_1x_4x_2x_3Suffix$	3
	Pref10000-Pref11000 *Pref10010-Pref11010*	$x_1x_2x_3x_4Suffix-x_3x_2x_1x_4Suffix-$ $x_2x_1x_4x_3Suffix-x_2x_1x_4x_3Suffix$	3
	Pref10001-Pref11011 *Pref10101-Pref11111*	$x_1x_2x_3x_4Suffix-x_2x_1x_3x_4Suffix-$ $x_4x_3x_2x_1Suffix$	2
4	*Pref10100-Pref11100* *Pref10110-Pref11110*	$x_1x_2x_3x_4Suffix-x_4x_3x_2x_1Suffix$	1
	Pref10011-Pref11001 *Pref10111-Pref11101*	$x_1x_2x_3x_4Suffix-x_4x_3x_2x_1Suffix-$ $x_2x_3x_4x_1Suffix$	2

Table 6.7. *Cases of embedding TQ_N into G_N, when the path is in the same G_4 of any super node*

6.3.5. *Embed3_edge(nodedep, nodarr) algorithm*

This procedure is used when $n>5$ and all paths are between two different G_4 in different super nodes.

Let $A = a_1a_2$ and $B = b_1b_2$, where $(a_1a_2, b_1b_2) = (00, 01), (00, 10), (01, 11),$ $(10, 11)$. For $n>5$, Embed3_edge(nodedep, nodarr) algorithm performs the different actions specified in the four stated following cases. Excepting the case when $n = 6$, *APref* is reduced to 0, *BPref* is reduced to 1.

For the case $n = 7$, $(APref, BPref) = (00, 01), (00, 10), (01, 11), (10, 11)$.

Twisted hypercube edge	Pancake path	Dilation
$APref00000$-$BPref00000$ $APref00010$-$BPref00010$	$x_1x_2x_3x_4x_5Suffix$-$x_3x_2x_1x_4x_5Suffix$	1
$APref00001$-$BPref00011$ $APref00101$-$BPref00111$	$x_1x_2x_3x_4x_5Suffix$-$x_2x_3x_1x_4x_5Suffix$- $x_5x_4x_1x_3x_2Suffix$-$x_4x_5x_1x_3x_2Suffix$- $x_1x_5x_4x_3x_2Suffix$	4
$APref00100$-$BPref00100$ $APref00110$-$BPref00110$	$x_1x_2x_3x_4x_5Suffix$-$x_3x_2x_1x_4x_5Suffix$- $x_5x_4x_1x_2x_3Suffix$-$x_1x_4x_5x_2x_3Suffix$	3
$APref00011$-$BPref00001$ $APref00111$-$BPref00101$	$x_1x_2x_3x_4x_5Suffix$-$x_2x_1x_3x_4x_5Suffix$- $x_5x_4x_3x_1x_2Suffix$-$x_3x_4x_5x_1x_2Suffix$- $x_1x_5x_4x_3x_2Suffix$	4
$APref01000$-$BPref01000$ $APref01010$-$BPref01010$	$x_1x_2x_3x_4x_5Suffix$-$x_4x_3x_2x_1x_5Suffix$- $x_5x_1x_2x_3x_4Suffix$	2
$APref01001$-$BPref01011$ $APref01101$-$BPref01111$	$x_1x_2x_3x_4x_5Suffix$-$x_2x_1x_3x_4x_5Suffix$- $x_4x_3x_1x_2x_5Suffix$-$x_5x_3x_1x_3x_4Suffix$- $x_2x_5x_1x_3x_4Suffix$	4
$APref01100$-$BPref01100$ $APref01110$-$BPref01110$	$x_1x_2x_3x_4x_5Suffix$-$x_4x_3x_2x_1x_5Suffix$- $x_5x_1x_2x_3x_4Suffix$-$x_4x_3x_2x_1x_5Suffix$	3
$APref01011$-$BPref01001$ $APref01111$-$BPref01101$	$x_1x_2x_3x_4x_5Suffix$-$x_2x_1x_3x_4x_5Suffix$- $x_4x_3x_1x_2x_5Suffix$-$x_5x_2x_1x_3x_4Suffix$- $x_3x_1x_2x_5x_4Suffix$	4
$APref10000$-$BPref10000$ $APref10010$-$BPref10010$	$x_1x_2x_3x_4x_5Suffix$-$x_3x_2x_1x_4x_5Suffix$- $x_5x_4x_1x_2x_3Suffix$-$x_1x_4x_5x_2x_3Suffix$	3
$APref10001$-$BPref10011$ $APref10101$-$BPref10111$	$x_1x_2x_3x_4x_5Suffix$-$x_2x_1x_3x_4x_5Suffix$- $x_5x_4x_3x_1x_2Suffix$-$x_3x_4x_5x_1x_2Suffix$- $x_4x_3x_5x_1x_2Suffix$-$x_4x_3x_5x_1x_2Suffix$	5
$APref10100$-$BPref10100$ $APref10110$-$BPref10110$	$x_1x_2x_3x_4x_5Suffix$-$x_5x_4x_3x_2x_1Suffix$	1
$APref10011$-$BPref10001$ $APref10111$-$BPref10101$	$x_1x_2x_3x_4x_5Suffix$-$x_5x_4x_3x_2x_1Suffix$- $x_4x_5x_3x_2x_1Suffixx_3x_5x_4x_2x_1Suffix$- $x_2x_4x_5x_3x_1Suffix$-$x_4x_2x_5x_3x_1Suffix$	5
$APref11000$-$BPref11000$ $APref11010$-$BPref11010$	$x_1x_2x_3x_4x_5Suffix$-$x_4x_3x_2x_1x_5Suffix$- $x_2x_3x_4x_1x_5Suffix$-$x_5x_1x_4x_3x_2Suffix$- $x_4x_1x_5x_3x_2Suffix$-$x_3x_5x_1x_4x_2Suffix$	5
$APref11001$-$BPref11011$ $APref11101$-$BPref11111$	$x_1x_2x_3x_4x_5Suffix$-$x_5x_4x_3x_2x_1Suffix$- $x_4x_5x_3x_2x_1Suffix$-$x_2x_3x_5x_4x_1Suffix$	3
$APref11100$-$BPref11100$ $APref11110$-$BPref11110$	$x_1x_2x_3x_4x_5Suffix$-$x_4x_3x_2x_1x_5Suffix$- $x_2x_3x_4x_1x_5Suffix$-$x_5x_1x_4x_3x_2Suffix$- $x_4x_1x_5x_3x_2Suffix$-$x_3x_5x_1x_4x_2Suffix$	5
$APref11011$-$BPref11001$ $APref11111$-$BPref11101$	$x_1x_2x_3x_4x_5Suffix$-$x_5x_4x_3x_2x_1Suffix$- $x_4x_5x_3x_2x_1Suffix$-$x_2x_3x_5x_4x_1Suffix$	3

Table 6.8. *Case 1 for A = 00 and B = 01*

For the sake of simplicity, cases 3 and 4 are not given in this chapter.

6.4. Dilations of many-to-one n-dimensional twisted hypercube embedded into n-dimensional pancake

6.4.1. *Lemma 6.1*

The n-dimensional twisted hypercube $TQ'_n = (V, U1)$ has many-to-one dilation 3 embedding into $G'_n = (P'_n, E'_n)$ for any $n>3$.

PROOF.– We prove lemma 6.1 by induction.

Twisted hypercube edge	Pancake path	Dilation
APref00000-BPref00000 *APref00010-BPref00010*	$x_1x_2x_3x_4x_5Suffix-x_3x_2x_1x_4x_5Suffix-$ $x_5x_4x_1x_2x_3Suffix$	2
APref00001-BPref00011 *APref00101-BPref00111*	$x_1x_2x_3x_4x_5Suffix-x_3x_2x_1x_4x_5Suffix-$ $x_5x_4x_1x_2x_3Suffix-x_1x_4x_5x_2x_3Suffix-$ $x_2x_5x_4x_1x_3Suffix$	4
APref00100-BPref00100 *APref00110-BPref00110*	$x_1x_2x_3x_4x_5Suffix-x_5x_4x_1x_2x_3Suffix-$ $x_1x_4x_5x_2x_3Suffix$	2
APref00011-BPref00001 *APref00111-BPref00101*	$x_1x_2x_3x_4x_5Suffix-x_5x_4x_3x_2x_1Suffix-$ $x_3x_4x_5x_2x_1Suffix-x_2x_5x_4x_3x_1Suffix-$ $x_4x_2x_5x_3x_1Suffix$	4
APref01000-BPref01000 *APref01010-BPref01010*	$x_1x_2x_3x_4x_5Suffix-x_4x_3x_2x_1x_5Suffix-$ $x_2x_3x_4x_5x_1Suffix-x_1x_5x_4x_3x_2Suffix-$ $x_3x_4x_5x_1x_2Suffix$	4
APref01001-BPref01011 *APref01101-BPref01111*	$x_1x_2x_3x_4x_5Suffix-x_5x_4x_3x_2x_1Suffix-$ $x_3x_4x_5x_2x_1Suffix-x_2x_5x_4x_3x_1Suffix$	3
APref01100-BPref01100 *APref01110-BPref01110*	$x_1x_2x_3x_4x_5Suffix-x_2x_1x_3x_4x_5Suffix-$ $x_5x_4x_3x_1x_2Suffix-x_1x_3x_4x_5x_2Suffix-$ $x_4x_3x_1x_5x_2Suffix-x_3x_4x_1x_5x_2Suffix$	5
APref01011-BPref01001 *APref01111-BPref01101*	$x_1x_2x_3x_4x_5Suffix-x_3x_2x_1x_4x_5Suffix-$ $x_5x_4x_1x_2x_3Suffix-x_2x_1x_4x_5x_3Suffix-$ $x_1x_2x_4x_5x_3Suffix-x_4x_5x_4x_1x_3Suffix$	3
APref10000-BPref10000 *APref10010-BPref10010*	$x_1x_2x_3x_4x_5Suffix-x_5x_4x_3x_2x_1Suffix-$ $x_3x_4x_5x_2x_3Suffix$	2
APref10001-BPref10011 *APref10101-BPref10111*	$x_1x_2x_3x_4x_5Suffix-x_2x_1x_3x_4x_5Suffix-$ $x_5x_4x_3x_1x_2Suffix-x_3x_4x_5x_1x_2Suffix-$ $x_4x_3x_5x_1x_2Suffix-x_5x_3x_4x_1x_2Suffix$	5

APref10100-BPref10100 *APref10110-BPref10110*	$x_1x_2x_3x_4x_5Suffix$-$x_3x_2x_1x_4x_5Suffix$- $x_5x_4x_1x_2x_3Suffix$	2
APref10011-BPref10001 *APref10111-BPref10101*	$x_1x_2x_3x_4x_5Suffix$-$x_2x_1x_3x_4x_5Suffix$- $x_5x_4x_3x_1x_2Suffix$-$x_3x_4x_5x_1x_2Suffix$- $x_1x_5x_4x_3x_2Suffix$-$x_5x_1x_4x_3x_2Suffix$	5
APref11000-BPref11000 *APref11010-BPref11010*	$x_1x_2x_3x_4x_5Suffix$-$x_4x_3x_2x_1x_5Suffix$- $x_5x_1x_2x_3x_4Suffix$-$x_2x_1x_5x_3x_4Suffix$- $x_3x_5x_1x_2x_4Suffix$	4
APref11001-BPref11011 *APref11101-BPref11111*	$x_1x_2x_3x_4x_5Suffix$-$x_3x_2x_1x_4x_5Suffix$- $x_4x_1x_2x_3x_5Suffix$-$x_5x_3x_2x_1x_4Suffix$- $x_2x_3x_5x_1x_4Suffix$	4
APref11100-BPref11100 *APref11110-BPref11110*	$x_1x_2x_3x_4x_5Suffix$-$x_4x_3x_2x_1x_5Suffix$- $x_5x_1x_2x_3x_4Suffix$-$x_2x_1x_5x_3x_4Suffix$- $x_3x_5x_1x_2x_4Suffix$	4
APref11011-BPref11001 *APref11111-BPref11101*	$x_1x_2x_3x_4x_5Suffix$-$x_4x_3x_2x_1x_5Suffix$- $x_5x_1x_2x_3x_4Suffix$	2

Table 6.9. *Case 2 for A = 00 and B = 10*

Base

For $n = 3$, Table 6.1 presents all paths between the embedded nodes of TQ_3 into G_3 with dilation 3.

Induction hypothesis

Suppose that for $k \leq n-1$, TQ'_{k-1} embedding many-to-one dilation 3 into G'_{k-1} is true. Let us now prove that it is true for $k = n$.

We have the following cases:

Case 1: k is even

$TQ'_n = (V, U1)$ is constructed by two copies of TQ'_{n-1}: one copy is prefixed by $0(0TQ'_{k-1})$ and the second one is prefixed by $1(1TQ'_{k-1})$. All nodes $A \in V$, such that, $A = 0Prefa_{n-3}a_{n-2}a_{n-1} = Pref_1a_{k-3}a_{k-2}a_{k-1}$, are embedded by Embed_node(A) algorithm, as shown in Table 6.4, into the first super node or into the projection $G'_k[k, k]$. All nodes $A \in V$, that is, $A = 1prefa_{k-3}a_{k-2}a_{k-1}$ or $A = Pref_2a_{k-3}a_{k-2}a_{k-1}$ are embedded into the second super node or into the projection $G'_k [k, 1]$, as shown in Table 6.5. That is to say, they are embedded into G'_{k-1}. However, the dilation of embedding into G'_{k-1} is 3 (hypothesis of induction).

Case 2: k is odd

Let $k = 2m+1$, where $m \in \mathbb{N}$, and TQ_n be obtained from two copies of $0TQ'_{2m}$ and $1TQ'_{2m}$, and suppose that for $N = 2m$ we have $0TQ'_N$ and $1TQ'_N$, that is to say, $00TQ'_{N-1}$, $01TQ'_{N-1}$ and $10TQ'_{N-1}$, $11TQ'_{N-1}$.

The Embed-node(A) algorithm, as shown in Table 6.1, embeds all nodes $A=00Prefa_{N-3}a_{N-2}a_{N-1}$ ($A \in V$) into the first super node or into the projection $G'_N [N, N]$, all nodes $A = 10Prefa_{N-3}a_{N-2}a_{N-1}$ into $G'_N [N, 1]$, all nodes $A = 01Prefa_{N-3}a_{N-2}a_{N-1}$ into $G'_N [N, 3]$, and all nodes $A = 11Prefa_{N-3}a_{N-2}a_{N-1}$ into $G'_N [N, 2]$.

In other words, we use only four super nodes among the k projections or super nodes. G'_N is a $(n-1)$-dimensional pancake graph embedding many-to-one dilation 3 into G'_N (hypothesis of induction).

6.4.2. Lemma 6.2

The n-dimensional twisted hypercube $TQ''_n = (V, U2)$ has many-to-one dilation 4 embedding into $G''_n = (P''_n, E''_n)$ for any $n>4$.

PROOF.– We use the same method to prove lemma 6.2, except that the embedding of the edges of TQ''_k is defined in Table 6.7 for the case where k is even and for the case where k is odd.

THEOREM 6.1.– The n-dimensional twisted hypercube $TQ_n = (V, U)$ has many-to-one dilation 5 embedding into $G_n = (P_n, E_n)$ for any $n>5$.

PROOF.– *Base*: For $n = 6$, Table 6.9 presents the case of different actions of embedding all edges of TQ_6 into G_6 with dilation 5.

For $n = 7$, Tables 6.8 and 6.9 and the data for cases 3 and 4 (not given) present the different actions of embedding all edges of TQ_7 into G_7 with dilation 5.

Induction hypothesis

Assume that this lemma 6.2 holds for $k \leq n-1$. That is, TQ_{k-1} embedding many-to-one dilation 5 into G_{k-1} is true.

Now we prove that this is true for $k = n$.

Case 1: k is even

There are two sub-cases:

– Case a:

As the twisted hypercube is defined as $TQ_k = (V, U)$, let A and $B \in V$, where $A = 0Prefa_{k-4}a_{k-3}a_{k-2}a_{k-1} = Pref_1a_{k-4}a_{k-3}a_{k-2}a_{k-1}$ as $Pref_1 = 0Pref$ and $B = Pref_1b_{k-4}b_{k-3}b_{k-2}b_{k-1}$. The embedding of $(A, B) \in U$ into the first super node or into the projection $G_k[k, k]$. All edges $(A, B) \in U$, such that, $A = 1prefa_{k-4}a_{k-3}a_{k-2}a_{k-1}$ or $A = Pref_2a_{k-4}a_{k-3}a_{k-2}a_{k-1}$, where $Pref_2 = 1Pref$, and the node $B = Pref_2b_{k-4}b_{k-3}b_{k-2}b_{k-1}$ are embedded into the second super node or into the projection $G_k[k, 1]$, in other words, into G_{k-1}. However, the dilation of embedding into G_{k-1} is 5 hypotheses of induction. □

– Case b:

As the twisted hypercube is defined as $TQ_k = (V, U)$, let A and B $\in$ V, $A = 0Prefa_{k-4}a_{k-3}a_{k-2}a_{k-1}$ or $A = Pref_1a_{k-4}a_{k-3}a_{k-2}a_{k-1}$ as $Pref_1 = 0Pref$ and B = $Pref_2b_{k-4}b_{k-3}b_{k-2}b_{k-1}$.

If we use Embed_node(A) algorithm, all nodes A are embedded into a super node $G_k[k, k]$ and all nodes B are embedded into a super node $G_k[k, 1]$. The different edges of TQ_k are embedded into different paths. The first node of every path is embedded into the super node $G_k[k, k]$, and the ending node is embedded into the super node $G_k[k, 1]$, that is to say, we use the different embedding edges outlined in cases 1, 2 and 3 (not given in this chapter) and case 4. In all cases, the dilation is 5.

Case 2: k is odd

There are two sub-cases:

– Case a:

Let $k = 2m+1$, where $m \in \mathbb{N}$, TQ_k is produced by two copies of $0TQ'_{2k}$ and $1TQ'_{2k}$. Suppose that for $N = 2k$ we have $0TQ'_n$, $1TQ'_N$, in other words, $00TQ'_{N-1}$, $01TQ'_{N-1}$, $10TQ'_{N-1}$ and $11TQ'_{N-1}$. Let A and $B \in V$, where $A = A_1A_2$, such that $A_1 = (00, 01, 10, 11)$, $A_2 = Prefa_{N-4}a_{N-3}a_{N-2}a_{N-1}$ as $Pref_1 = A_1Pref$; hence, $A = Pref_1a_{N-4}a_{N-3}a_{N-2}a_{N-1}$ and $B = Pref_1b_{N-3}b_{N-2}b_{N-1}b_N$. The embedding of $(A, B) \in U$ is into the first super node $G_N[N, N]$ if $A_1 = 00$, into the second super node $G_N[N, 1]$ if $A_1 = 10$, into the third super node $G_N[N, 3]$ if $A_1 = 01$, and into the fourth super node $G_N[N, 2]$ if $A_1 = 11$. The dilation in all super nodes is 5 (hypothesis induction).

– Case b:

Let A and $B \in V$, and $A = A_1A_2$, $B = B_1B_2$ as $(A_1, B_1) = (00, 01)$, $(00, 10)$, $(01, 11)$, $(10, 11)$ and $A_2 = Pref_1a_{N-4}a_{N-3}a_{N-2}a_{N-1}$, $B_2 = Pref_1b_{N-3}b_{N-2}b_{N-1}b_N$.

The embedding of $(A, B) \in U$ is into a different path between two super nodes $(G_N[N, N], G_N[N, 3])$, $(G_N[N, N], G_N[N, 1])$, $(G_N[N, 3], G_N[N, 2])$, $(G_N[N, 1], G_N[N, 2])$. Each super node contains exactly 2^{l-1} G_4. In other words, case 1 or case 2 is used, because the first node of the different paths is located in one node of G_4 of the super node $G_N[N, N]$, and the ending node is located in one node of G_4 of the super node $G_N[N, 3]$. Or for all edges of TQ_N having the first extremity a node prefixed by $00Pref$, and the second extremity a node prefixed by $01Pref$, for instance, cases 1, 2, 3 and 4 (cases 3 and 4 are not given in this chapter) are used. In all cases, the dilation is 5.

6.5. Conclusion

It is both practically significant and theoretically interesting to investigate the embeddability of different architecture into pancake. The main purpose of this chapter is the many-to-one 5 dilation embedding of an n-dimensional twisted hypercube into a pancake of n dimensions. The study of the dilation of this new many-to-one embedding function is explained in three steps. The first step is embedding many-to-one dilation 3 of all edges in paths in the same G_3 components of a super node, as proved by lemma 6.1. The second step is that, for all paths, results of many-to-one dilation 4 embedding graph are in the same G_4 components of a super node; in other words, the path is between two G_3 of the same G_4, as proved by lemma 6.2, and the latter step is the general embedding many-to-one dilation 5 of all edges of the n-dimensional twisted hypercube TQ_n in the paths between two different super nodes.

In the feature of this work, it is more interesting to study the one-to-one embedding case and the fault-tolerant embedding of an n-dimensional twisted hypercube and the crossed hypercube into an n-dimensional pancake graph.

6.6. Bibliography

[ABR 91] ABRAHAM S., PADMANABHAN K., "Twisted cube: a study in asymmetry", *Parallel Distributions and Computer*, vol. 13, pp. 104–110, 1991.

[AKE 89] AKERS S.B., Krishnamurthy B., "A group-theoretic model for symmetric interconnection networks," *IEEE Transactions on Computers*, vol. 4, pp. 555–566, 1989.

[BOU 98] BOUABDALLAH A., HEYDEMANN M.C., OPATRNY J. *et al.*, "Embedding complete binary trees into star and pancake graphs", *Theory Computer System Journal*, vol. 31, no. 3, pp. 279–305, 1998.

[CHA 99] CHANG C.P., WANG J.N., HSU L.H., "Topological properties of twisted cubes", *Information Sciences*, pp. 147–167, 1999.

[FAN 02] FAN J., "Hamilton-connectivity and cycle-embedding of the Mobius cubes", *Information Processing Letters*, vol. 82, pp. 113–117, 2002.

[FAN 00] FANG W.C., HSU C.C., "On the fault-tolerant embedding of complete binary trees in the pancake graph interconnection network", *Information Sciences*, vol. 126, nos. 1–4, pp. 191–204, 2000.

[FEM 17] FEMMAM S., *Signals and Control Systems: Application for Home Health Monitoring*, ISTE, London and John Wiley & Sons, New York, 2017.

[HEY 97] HEYDARI M.H., SUDBOROUGH I.H., "On the diameter of Pancake network", *Algorithms Journal*, vol. 25, no. 1, pp. 67–94, 1997.

[HSI 98] HSIEH S.Y., CHEN G.H., HO C.W., "Embed longest rings onto star graphs with vertex faults", *International Conference on Parallel Processing, Proceeding*, pp. 140–147, 1998.

[HSI 99] HSIEH S.Y., CHEN G.H., HO C.W., "Fault-free Hamilton-cycles in faulty arrangement graphs", *IEEE Transactions on Parallel Distribution System*, vol. 10, pp. 223–237, 1999.

[HUA 02] HUAN W.T., TAN J.M., HUNG C.N. *et al.*, "Fault-tolerant Hamiltonocity of twisted cubes", *Parallel Distributions and Computer*, vol. 62, pp. 582–604, 2002.

[HUN 02] HUNG C., LIANG K., HSU L.-H., "Embedding Hamiltonian paths and Hamiltonian cycles in faulty pancake graphs", *International Symposium on Parallel Architectures, Algorithms and Networks (ISPAN '02)*, pp. 157–162, 2002.

[HUN 03] HUNG C., HSU H., LIANG K. *et al.*, "Ring embedding in faulty pancake graphs", *Information Processing Letters*, vol. 86, pp. 271–275, 2003.

[HWA 00] HWANG S.C., CHEN G.H., "Cycles in butterfly graphs", *Networks Journal*, vol. 35, pp. 161–171, 2000.

[KAN 95] KANEVSKY A., FENG C., "On the embedding of cycles in pancake graphs", *Parallel Computer Journal*, vol. 21, no. 6, pp. 923–926, 1995.

[KUL 95] KULASINGHE P., BETTAYEB S., "Multiply-twisted hypercube with five or more dimensions is not vertex-transitive", *Information Processing Letters*, vol. 5, pp. 33–36, 1995.

[LIN 08] LIN C.-K., TAN J.J.M., HUANG H.-M. *et al.*, "Mutually independent Hamiltonianicity of pancake graphs and star graphs", *The International Symposium on Parallel Architectures, Algorithms, and Networks (i-span)*, pp. 151–158, 2008.

[LIN 10] LIN J.-C., YANG J.-S., HSU C.-C. *et al.*, "Independent spanning trees vs edge-disjoint spanning trees in locally twisted cubes", *Information Processing Letters Journal*, vol. 110, no. 10, pp. 414–419, 2010.

[MEN 92] MENN A., SOMANI A.K., "An efficient sorting algorithm for the star graph interconnection network", *19th International Conference In Parallel Computation, Proceeding*, vol. 10, pp. 3–20, 1992.

[MIL 94] MILLER Z., PRITIKIN D., SUDBOROUGH I., "Near embedding of hypercubes into Cayley graphs on the the symmetric group", *IEEE Transactions on Computers*, vol. 43, pp. 13–22, 1994.

[MOR 96] MORALES L., SUDBOROUGH I., "Comparing star and pancake networks", *Symposium Parallel and Distributed Processing (SPDP)*, pp. 612–615, 1996.

[QIU 91] QIU K., AKL S.G., STOJMENOVIC I., "Data communication and computational geometry on the star interconnection networks", *Proceedings of the 3rd IEEE Symposium on Parallel and Distributed Processing*, pp. 125–129, 1991.

[QIU 92] QIU K., The star and Pancake interconnection networks: proprieties and algorithms, PhD Thesis, Queens University, Canada, August, 1992.

[ROW 93] ROWLEY R.A., BOSE B., "On ring embedding in de-Bruijn networks", *IEEE Transaction on Computer Journal*, vol. 12, pp. 1480–1486, 1993.

[ROW 98] ROWLEY R.A., BOSE B., "Fault-tolerant ring embedding in de Bruijn networks", *IEEE Transactions on Computer*, vol. 12, pp. 1480–1486, 1998.

[SEN 03] SENGUPTA A., "On ring embedding in Hypercubes with faulty nodes and links", *Information Processing Letters*, vol. 68, pp. 207–214, 2003.

[SEN 97] SENOUSSI H., LAVAULT C., "Embeddings into the pancake interconnection network", *High Performance Computing and Grid in Asia Pacific Region, International Conference on High-Performance Computing on the Information Superhighway, HPC–Asia '97*, pp. 73–78, 1997.

[XU 10] XU M., "Edge-pancyclicity and Hamiltonian connectivity of twisted cubes", *Acta Mathematica Sinica*, vol. 26, no. 7, pp. 1315–1322, 2010.

List of Authors

Raymond ASCHHEIM
Quantum Gravity Research
Laboratory
Los Angeles, California
USA

Mohamed Ikbal BENAKILA
Computer Consultant
Paris
France

Sébastien BINDEL
University of Upper Alsace
Mulhouse
France

Frédéric DROUHIN
University of Upper Alsace
Mulhouse
France

Smain FEMMAM
University of Upper Alsace
Mulhouse
France

Laurent GEORGE
University of Paris-Est
LIGM, ESIEE
Paris
France

Marc GILG
University of Upper Alsace
Mulhouse
France

Xiaoting LI
ECE Paris
Paris
France

Faouzi M. ZERARKA
University of Biskra
Biskra
Algeria